NOTES FROM THE HOLOCENE

NOTES FROM THE HOLOCENE

A BRIEF HISTORY OF THE FUTURE

DORION SAGAN

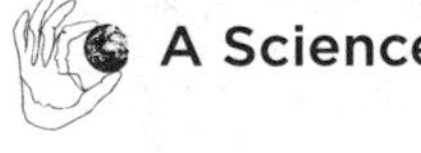

A Sciencewriters Book

CHELSEA GREEN PUBLISHING COMPANY
WHITE RIVER JUNCTION, VERMONT

 **A Sciencewriters Book**

scientific knowledge through enchantment
Sciencewriters Books is an imprint of Chelsea Green Publishing. Founded and codirected by Lynn Margulis and Dorion Sagan, Sciencewriters is an educational partnership devoted to advancing science through enchantment in the form of the finest possible books, videos, and other media.

Developmental Editor: John Barstow
Project Manager: Collette Leonard
Copy Editor: Laura Jorstad
Proofreader: Robin Catalano
Indexer: Pam Rider
Designer: Peter Holm, Sterling Hill Productions
Cover Design: Rick Pracher

Printed in the United States of America
First printing, August 2007
10 9 8 7 6 5 4 3 2 1

Library of Congress Cataloging-in-Publication Data
Sagan, Dorion, 1959-
Notes from the Holocene : a brief history of the future / Dorion Sagan.
p. cm.
"A Sciencewriters Book."
Includes index.
ISBN 978-1-933392-32-5
1. Cosmogony. 2. Earth. 3. Biosphere. 4. Science—Philosophy. I.Title.

QE506.S24 2007
500—dc22

2007022115

Our Commitment to Green Publishing
Chelsea Green sees publishing as a tool for cultural change and ecological stewardship. We strive to align our book manufacturing practices with our editorial mission and to reduce the impact of our business enterprise on the environment. We print our books and catalogs on chlorine-free recycled paper, using soy-based inks whenever possible. This book may cost slightly more because we use recycled paper, and we hope you'll agree that it's worth it. Chelsea Green is a member of the Green Press Initiative (www.greenpressinitiative.org), a nonprofit coalition of publishers, manufacturers, and authors working to protect the world's endangered forests and conserve natural resources.

Notes from the Holocene was printed on Ecobook 50 Natural, a 50 percent post-consumer-waste recycled, old-growth-forest-free paper supplied by Maple-Vail.

Chelsea Green Publishing
Post Office Box 428
White River Junction, Vermont 05001
(802) 295-6300
www.chelseagreen.com

— CONTENTS —

This book contains wild speculations.
Read at your own risk.*

***This statement has not been verified by the FDA, MDA, USDA, APA, DEA, GSA, CIA, NSA, AA, AAA, or AAAA.**

EARTH

> Yes, happily language is a thing: it is a written thing, a bit of bark, a sliver of rock, a fragment of clay in which the reality of the earth continues to exist.
>
> —Maurice Blanchot

Where are we
going? What are we doing on this planet? Why—if that's the right question—are we here? The question "Why are we alive?" was, according to a 2006 survey by the Chinese search engine Baidu, the second most often typed *why* question, right between "Why did they go on the Long March?" and "Why do we need to drink water?" In this book I use physics, evolutionary history, science fiction, knowledge of magic tricks, and even a little metaphysics to speculate on basic questions of who and what we are in relationship to the Earth and the universe. To anticipate, I conclude that life has a physical purpose, that there are many universes, that the Earth is an organized system that may be conscious, and that it has already begun a process of reproduction that may take its offspring—including, perhaps, us—to the stars. I conclude that linear time and free will are possibly illusions, and that the typical notion of God is hopelessly naive. I use a professional knowledge of sleight-of-hand magic to treat real mysteries such as where consciousness comes from and the reasons we seem both so impossibly unlikely and yet so full of imperfections. I think you'll be surprised at the simplicity of some of the probable answers.

What are the prospects for humanity? How can unfeeling particles give rise to feeling beings? Who are we from the broad vantage point of deep time, in which not just primates but microbes preceded us, and God knows what will descend from or replace us? Is the biosphere imperiled? Are we? Can it, or us, be saved? How? What is the future of a biosphere that is more than four billion years old—approximately a thousand times older than the human species, and two million times older than the oldest cities?

I am interested in provisional answers to some of the deepest, most persistent questions we can ask ourselves. I look to science but also to the speculations of science fiction, the revelations of mystics, and the logic of philosophers. What are the most radical ways we can see ourselves that might possibly be true? What is the nature of ultimate reality? Of our odd, orderly universe? I will address these cosmic questions, but let's start with the most basic thing—the Earth, the living part of which is known as the biosphere.

The very word *biosphere* is subtly subversive, undermining the nationalistic mind-set. Mere mention of it provides a clue that our ultimate allegiance is beyond politics and race. The very substance of our bodies comes from and will return to the biosphere when we die. We depend on other species, many of whose roles in maintaining the biosphere fit for life are more important than our own. Indeed, our rampant growth has perturbed the planet as a whole, arguably compromising its functioning in a way that will imperil and perhaps eliminate us.

On Earth there is no escape, no exit, from global ecology. We may ignore it but it is still there, like the truth. Still, we now know, maybe not enough, yet more than our ancestors did. Thanks to science, especially planetary exploration, we can look at ourselves in a more objective light. Just as the best part of a journey can be the new perspective it gives us on our life, so the greatest windfall of the space program is the new appreciation it gives us of our planetary home. Astronauts report that when in orbit, going around the Earth every forty-five minutes, time and space—or rather, our ordinary perspective on time and space—are radically disturbed. There is no day or night, no up or down. Sunrise follows sunset very quickly. In orbit, our star's light cuts through the thin ribbon of the atmosphere, illuminating the interior of the capsule with all the colors of the rainbow. Then, forty-five minutes later, night falls. Earth (eclipsing the sun) becomes the place where there are no stars.

The concept dawns that we belong to the biosphere, and not just the biosphere of the Earth but of the cosmos, as well. The nationless, mapless perspective of Earth seen by an astronaut in space—in which human beings are visible, if at all, only by the lights of our cities at night—puts us in our place. The new views of space, time, and home suggest that our usual perspectives are just that, perspectives. The way things are, the way we see them, depends on where we are and how we see. And those things can change.

"In what sense 'is there,'" writes Italian philosopher Maurizio Ferraris, "a star that exploded a thousand years ago, and that we see now? It is to be noted that, according to the distinction between phenomenon and noumenon, everything visible—ourselves included—could be nothing but memory and phenomenalization, no less than stars that have exploded, and appeared precisely when they have ceased to be noumena." Ferraris here refers to the philosophical distinction (first made by Immanuel Kant) between the way things are in themselves, which we do not know directly, and the way things come to us—necessarily altered—through our senses and brains. The thing-in-itself, *das Ding an sich* in Kant's native German, is the noumenon; how something appears to us in our perception, including our perceptions aided by scientific instruments, is the phenomenon.

The question *Why are we here?—Why are we alive?*—admits of various answers. It may be because of God. It may be because of chance. Or, as Ferraris suggests, we may not even be here. Everything may already have happened, and we may be an aftereffect of the type that occurs when a supernova explodes but the light takes time to reach our eyes.

Now you see it, now you don't.

The view of Earth from the moon
known as *Earthrise,* taken December 24, 1968, seven months after the first lunar landing—first a black-and-white photograph by *Apollo 8* commander Frank Borman and then, two minutes later, two color photos by astronaut William A. Anders—is an epiphany. For the first time we see the place where we all live. And it looks like more than a place: it looks vibrant, alive, a being in its own right. The face of the Earth is our own.

Like Narcissus we at first do not recognize what we are seeing. And NASA—or the politicos and bureaucrats who control it—don't seem to have gotten this point. From 2002 through 2006 NASA's mission statement read: "To understand and protect our home planet; to explore the universe and search for life; to inspire the next generation of explorers as only NASA can." No more. As reported on the first page of *The New York Times* in the July 22, 2006, Saturday edition, the opening phrase *To understand and protect our home planet* was eliminated. Was somebody threatened by that line? Did they think it presented a false promise? That NASA's charter was to understand other planets but not our own?

It is ironic. If I could be around to collect, I would bet that the future will regard our newfound understanding of our planet as *the* greatest scientific success of the space program. James Lovelock, the English atmospheric chemist and inventor who shared an office with my father at NASA's Jet Propulsion Laboratory and was hired by the lab to design instrumentation to detect life on Mars, realized in the 1970s that the red planet was the dead planet. Its atmosphere lacked the continuous presence of co-reactive gases such as those that mark our planet. These various gases—which include methane, ammonia, nitrous oxide, nitrogen as a gas, and oxygen—together should react with each other, being nearly indetectible. Yet they stay around in our atmosphere. Together they show that something complex is occurring at our planetary surface. In various concentrations the chemicals we find in ourselves, in other life forms, and floating in the air are found in some cases in concentrations thousands of times higher than would be expected given the rules of standard chemical mixing. The complex system is not just at the level of the cell or the animal, but the planet. Given Earth's oceans, for example, nitrogen, the most common element found in our atmosphere, should be found in a more stable form as a nitrate ion dissolved in the sea. Chemically out of whack, that the gases exist simultaneously in each other's presence is a planetary sign. It suggests that some local energy-dependent organizing process is actively generating entropy into space to compensate for the unexpected far-from-random complexity at the surface of our planet. This process turns out to be life itself, which, continuously producing reactive compounds, expands its energy-storing, energy-using, intelligence-gathering, form-multiplying realm.

Unlike Narcissus who, in the Greek myth, accompanied by audio courtesy of the Greek nymph Echo (who repeats Narcissus' phrases), falls in love with his image and drowns, we—as attested by the recent restatement of the NASA mission statement—have, at least institutionally, turned away from this image, *Earthrise,* which nature photographer Galen Rowell called "the most influential environmental photograph ever taken." If Narcissus had too much self-love for his face, and drowned because of it, we have too little respect for ours and may burn—a natural, predictable consequence, rather than supernatural punishment—for our environmental sins. It may even be too late for us to stop the global heating fostered by our rampant reproduction and industrialization of Earth's surface, as well as our collective ignorance—now institutionally encouraged—of the global system in which we are embedded.

As it turns out

there is plenty of evidence for what at first seems a bizarre science-fiction idea—that Earth is a giant living being, perhaps a superorganism as far beyond us as we are beyond our constituent cells.

Careful study reveals that the chemistry and temperature of the atmosphere, oceans, and sediments—the surface of the Earth—is under active control. The data are much like those seen for temperature regulation in a warm-blooded mammal. For example, we have evidence of liquid water existing on Earth's surface for billions of years, despite nuclear physics suggesting that the sun had 30 percent less luminosity when it was young, some five billion years ago. In other words, Earth's surface, full of life, has managed to cool itself to counter the increased output from the sun, which might otherwise have scorched Earth's living surface to a crisp. So, too, although oxygen is an extremely reactive gas—in liquid form with hydrogen, its controlled reaction fires rockets into space—it continues to account for approximately one-fifth of Earth's atmosphere. It has done so for the last 500-million-odd years. According to the standard rules of chemical mixing, this should not happen—just as, according to mathematical calculations of random particle interactions, a roughly symmetrical being with fingernails and hair such as yourself should not be here. Religionists would tend to say that your presence is a miracle, testimony to God. Scientists tend to say that there is nothing miraculous about it; that you are the result of billions of years of natural selection. However, as Richard Dawkins has pointed out, there is only one Earth with no evidence, as there is normally in evolution, of a bunch of variants that died out. So initially at least, it is difficult to see where the biosphere's ability to thermoregulate and maintain its surface chemistry comes from.

Nonetheless, it is there. The mean temperature of the lower atmosphere is about twenty-two degrees Celsius; the pH of the planetary surface averages just over 8.0; and chemical evidence from the fossil record shows that such anomalies have persisted for many millions of years. Astrophysical theories of the birth and death of stars uniformly suggest that the sun began five billion years ago, smaller than it is now; it progressively increased in size and luminosity. Yet fossil evidence shows that liquid water, including rivers and lakes, existed at the Earth's surface for the last three billion years. Did the biosphere steady its temperature as the sun grew in luminosity and the radiation reaching the Earth increased?

The traditional explanation for such temperature control assumes that at best life played an incidental role. Sometimes called the Goldilocks problem, the question is why Venus became too hot and Mars too cold for life while the Earth remained "just right." One answer to this climatological problem is that these three planets differ in their respective abilities to cycle carbon dioxide between their oceans and atmospheres. Mars, in part because it was relatively small, lost its internal heat, becoming so cold that it could no longer release carbon dioxide from carbonated sediments to replace carbon dioxide, leaving the atmosphere in the form of carbonic acid rainfall. Since carbon dioxide was no longer available to let in sunlight and trap it as heat, Mars went into a deep freeze. On Venus, by contrast, there was so much carbon dioxide heating up the planet as infrared radiation that a runaway greenhouse effect arose, creating a hellish planet too hot for life. From this traditional viewpoint, life's happy climate is incidental, a lucky accident of a temperature-regulating carbon dioxide cycle between crust and atmosphere, which was established on Earth but disrupted on Mars and Venus. Life took advantage of, but had nothing to do with, the temperature of the Earth being "just right" for billions of years despite the increasing heat of the sun.

From another perspective, however, life actively participated in climatological modulation. Many animals regulate their temperature. All animals are evolved multicellular clones. Might not the Earth, as a superorganism containing organisms, regulate itself by feedback processes similar to those at work in a human being made of cells? Honeybees act together to modulate the temperature of their hive and are recognized, along with other social insects, as forming superorganisms. As a single giant being, the Earth, too, may regulate its temperature, chemical composition, and environment in general. This may be nothing mystical but rather the gestalt result of the organisms busily growing within and at Earth's surface. Thermodynamically, organisms are not isolated or closed but open systems: They are continuously exchanging matter and energy with their environment. Since they are open, what they do will tend to affect the biosphere as a whole. Moreover, since the biosphere's organisms grow only within certain temperature parameters, and within other natural constraints, the sum of their behavior will confer order upon (will organize) the global environment as a whole. This may seem counterintuitive because we don't normally think of the organism and the environment as part of the same thing. But they are. Organisms not only continuously exchange matter—taking in and putting out liquids, solids, and gases—with their surroundings, but grow while doing so. This has the effect of profoundly altering the surroundings in the shape of life, as much as or even more than life adapts to the environment.

Some very good conservative scientists at first denied the evidence for physiology-like control of temperature, atmospheric chemistry, ocean salinity, and other variables at Earth's surface—not because it was the scientific thing to do, but because such control, at a planetary level, tended to remind them of mental processes that were too much for them to accept as belonging to a planet. To address the concerns of scientists who could not see how, in the absence of natural selection, the biosphere itself might behave as an organism, Lovelock and co-worker Andrew Watson made a mathematical model called Daisy World. This is a computer-generated planet containing black and white daisies that can control its temperature to keep cool despite the increasing luminosity of a nearby sun. Like the real biosphere (which has survived damage from impacts equivalent to third-degree burns of 60 percent of the skin area of a human fire victim), the model biosphere recovers.

Daisy World, a breathtaking revision of the entire field of theoretical ecology, shows that no mystical assumptions need be made for organisms to control features of their environment as a unity: Planetary regulation in the model results quite naturally from the effects on the environment of individually growing organisms. As the digital sun increases in temperature, white daisies, reflecting light, cool the planet; the cooler it gets, however, the more black daisies, whose color allows them to absorb heat, grow, thereby warming the planet. While just one color of daisy can regulate global temperature, black and white together do a better job of it. And notice that the temperature regulation is not due to natural selection; it is, rather, an offshoot of the flowers' reflectivity (albedo) combined with their properties of growing and reproducing only within certain temperatures.

Since the original, relatively simple Daisy World model, British ecologist Stephan Harding and others have made more complex models that include other species such as cows that eat daisies, predators that eat the cows, and so on. Natural selection has been introduced in some of the models, but strictly speaking it is not required to control planetary temperature. The temperature regulation of Daisy World comes about through the activity of organisms growing within certain temperatures. The mysterious complexity—which some have dismissed because it seems to suggest some sort of conscious control of an entire planet that would seem impossible for "lower" organisms—arises quite naturally from their growth properties. In fact, what we call consciousness may be an offshoot of a far more widespread unconscious physiology. Planetary regulation of environmental variables need not require forethought or consciousness. Rather than view the organism-like features of Earth's surface as unacceptably mental, should we perhaps consider thought intrinsically physiological?

On our real Earth temperature regulation is achieved by nondaisy means. The main actors may be microbes. To protect themselves, for example, some plankton (called coccolithophores) produce dimethyl sulfide, a gas implicated in the genesis of marine stratus clouds. When it gets hot, the plankton grow and produce their waste gas, which reacts in the atmosphere to become cloud condensation nuclei—that is, the material substrate for forming raindrops. Other things being equal, if these plankton grow more when it is hotter, they produce more of this gas, leading to more cloud cover and thus the reflection of light and heat into space. This process, although not as simple and straightforward as the daisies, may well act as a real-world natural thermostat, cooling the planet or sections thereof. An article in *Nature* contested this supposed mechanism of planetary physiology by noting that sulfur dioxide emissions, produced mainly by industry in the developed countries, have not noticeably increased cloud cover over the Northern Hemisphere or led to a noticeable decrease in Northern Hemisphere temperatures within the last hundred years. The aerosols from industrial emissions may be indeed cooling the Northern Hemisphere as expected, but the effect is masked by well-documented global warming. The aerosols produced by industry, in other words, lead to cooling cloud cover—but this is more than made up for by human-produced carbon dioxide emissions. Indeed, the notion that sulfur aerosols temporarily cool the planet in the face of an opposite trend of global warming is strongly suggested by the European heat wave in 2003, which killed thirty thousand people: At precisely this time government regulations kicked in to reduce industrial sulfur emissions to combat acid rain. Adding environmental insult to economic injury is the possibility that global recession, reducing industry and therefore aerosols, could cause a massive, unintended spike in global mean temperature.

Sulfur dioxide
is not dimethyl sulfide but is assumed to work similarly. Whether or not sulfur-emitting algae regulate global temperature, real temperature regulation of the planet's surface is more likely to be accomplished by production and consumption of carbon dioxide, methane, water vapor, and other "greenhouse" gases than by the simple color changes of the Daisy World model. Another interesting mechanism is aromatic plant compounds called terpenes; these essential-oil-containing molecules have been implicated in raindrop formation. Bacteria, growing on leaves, may also be involved. Still another probability are forests themselves, which people cut down to make room for farmland, for lumber, or to make grassland for grazing cows. Loggers will tell you that when trees are cut down, the surface becomes noticeably warmer. That's because trees, especially rain forests, not only make shade but also help create sunlight-reflecting clouds. Together the world's forests may be an underrated major means of planetary cooling that has helped keep Earth's temperature "just right" despite the increasing luminosity of the sun. Space satellite observations show that the areas over rain forests in Borneo, the Amazon, and elsewhere are as cool as Siberia in winter. The water vapor that flows through trees, through the openings or stomata of their leaves to re-condense and fall as rain, certainly seems to be a major mechanism of global cooling. We cut down trees at our own peril. Locally, removal of trees leads immediately to the aforementioned temperature spikes at ground level. The English idiom *not out of the woods yet,* meant to indicate a process that is as yet incomplete, is backward from an ecological standpoint: The woods are part of the mechanism of climatological cooling that protects us. They are also the native environment for our species, the environment in which our primate ancestors evolved. Woe to our memory and global health should we ever truly and finally get out of the woods.

The *Nature* article shows that the notion of the Earth's physiology is based on complex interactions among organisms with their environment. Although the main players are not flowers, and reflectivity is only part of global temperature control, the Daisy World model shows in principle how the globally distributed organic thermostat works.

Moreover, the Daisy World model sheds light on the intriguing question of what constitutes consciousness. For the net effect of the daisies is to create an appearance of global intelligent behavior. This is, of course, without any Internet, technology as such, or "advanced" human beings. The daisies in the model behave as if they knew what they were doing. The net result of their behavior is similar to that analyzed by Adam Smith under the rubric of "the invisible hand": Each organism, acting for itself, confers organization on the whole. Although we associate such thermoregulation with conscious engineering, it is of course natural. We may consciously decide to put on our coat if cold, or go to the store and buy a fan or air conditioner if too hot. But we also shiver and sweat to thermoregulate without conscious forethought. Physiology seems to reflect a natural, unconscious intelligence based on feedback.

Consciousness itself may have grown out of this "wisdom of the body." Rather than be mystified by the conscious-like behavior of a planet able to regulate its temperature, we should appreciate that our intelligence is probably natural, the evolutionary outgrowth of ecological processes. The great Montaigne-like essayist, medical doctor, cigarette smoker, and director of the Sloan-Kettering Institute, Lewis Thomas, wrote this in commemoration of the twentieth anniversary of the moon landing:

> But the moment that really mattered came later, after the equipment had been set up for taking pictures afield. There, before our eyes, causing a quick drawing in of breath the instant it appeared on television screens everywhere, was that photograph of Earth. . . . [I]t was the loveliest object human beings had ever looked upon: home.
>
> Moreover, as anyone could plainly see in the photograph, it was alive. That astonishing round thing, hanging there all alone, totally unlike any of the numberless, glistening but dead-white works of geometry elsewhere in space, was a living thing, a being . . . exploding with meaning. . . . It maintains in precision the salinity and acid-base balance of its oceans, holds constant over millions of years the exactly

> equilibrated components of its atmosphere with the levels of oxygen and carbon dioxide at just the optimal levels for respiration and photosynthesis. It lives off the sun, taking in the energy it requires for its life and reflecting away the rest into the unfillable sink of space. . . . The one biological function the Earth does not yet perform to qualify for the formal definition of an organism is reproduction. But wait around, and keep an eye on it. In real life, this may turn out to be what it started to do 20 years ago. . . . the Earth may be entering the first stages of replication, scattering seeds of itself, perhaps in the form of microorganisms similar to those dominating the planet's own first life for the first two billion years of its Precambrian period. . . . Finally, as something to think about, there is the strangest of all paradoxes: the notion that an organism so immense and complex, with so many interconnected and communicating central nervous systems at work, from crickets and fireflies to philosophers, should be itself mindless. I cannot believe it.

Life's effect upon the Earth's crust is so pervasive that even the great tectonic movements beneath the oceans may be connected to, if not prompted by, organismic collectives. Though not yet proven, a series of possible links suggests that the most apparently inanimate and Herculean geological processes may be under biological dominion. Life behaves, as the groundbreaking (so to speak) Russian-Ukranian scientist Vladimir Vernadksy (1863–1945) put it, "as a geological force."

To glimpse how life may be implicated in plate tectonics, consider first that the amount of salt we might expect to find in the ocean—ten times existing levels—isn't there: Chemical and physical calculations show that salt from rivers and streams should pile up cumulatively in the great marine basins. Since there is no geological means to remove this salt, we have to assume that life itself keeps the oceans free from the high levels of salt that would otherwise accumulate there. Marine creatures cannot tolerate high salt concentrations; they are poisoned, not tickled but pickled to death, by such levels. So somehow the salt is removed. But if living beings take the salt, what do they do with it? Where do they put it? And how?

The missing salt appears to lie at the edges of the sea, dried up in coastal deposits called evaporites. These are sand-ridden expanses—flats and fields rich in minerals and salts. Charles Darwin cast his curious eyes over such deposits; we find references to them in his notebooks. We now know that what Darwin saw were packed communities of bacteria. The exudates, or cellular secretions, of these beings wrap around particles of salt, preventing them from being dissolved again into the water. Such natural salt mining might work to hold vast quantities back from the ocean, but only if the land is broad and flat and in the tropics or subtropics so that sufficient evaporation can take place. With limestone acting as a lubricant, landmasses slide over the mantle, slipping especially over stationary sources of tremendous heat coming from within the Earth. The mobile plates tend to accumulate in the tropical and semitropical zones where the slippery limestone is produced. Slipping about, they may create platforms for salt removal in their wake. While the details of sifting salt from the oceans remain unclear, ocean salinity is probably being regulated. Without obvious foresight, and by means of the immense continuous growth of microorganisms able to coat salt and thus prevent it from returning to solution in microbial mats and crystalline lagoons, life maintains its environment, and the seas remain habitable environments rather than supersaturated vats of poison, suitable for salt curing or pickling but not coccolithophores and whales and octopuses and other marine creatures.

According to Don L. Anderson, the Eleanor and John R. McMillan Professor of Geophysics, Emeritus, at the California Institute of Technology, the conversion of atmospheric carbon dioxide into calcium carbonate sediment—a chemical reaction that on Earth takes place primarily thanks to life—could have destabilized the rocks of Earth's crust so much that it began the process of plate tectonics: "[P]late tectonics may exist on the Earth because limestone-generating life evolved here."

The idea that Earth is alive is not new. Native Americans held people to be literal embodiments of the land. Jorge Luis Borges, known for his scholarly fables and mysterious essays, informs us in his *Book of Imaginary Beings*:

> The sphere is the most uniform of solid bodies since every point on its surface is equidistant from its centre. Because of this, and because of its ability to revolve on an axis without straying from a fixed place, Plato (*Timaeus,* 33) approved the judgment of the Demiurge, who gave the world a spherical shape. Plato thought the world to be a living being and in the *Laws* (898) stated that the planets and stars were living as well. In this way, he enriched fantastic zoology with vast spherical animals and cast aspersions on those slow-witted astronomers who failed to understand that the circular course of heavenly bodies was voluntary. . . . During the Renaissance, the idea of Heaven as an animal reappeared in Lucilio Vanini; the Neoplatonist Marsilio Ficino spoke of the hair, teeth, and bones of the earth; and Giordano Bruno felt that the planets were great peaceful animals, warm-blooded, with regular habits, and endowed with reason. At the beginning of the seventeenth century, the German astronomer Johannes Kepler debated with the English mystic Robert Fludd which of them had first conceived the notion of the earth as a living monster, "whose whale like breathing, changing with sleep and wakefulness, produces the ebb and flow of the sea." The anatomy, the feeding habits, the colour, the memory, and the imaginative and shaping faculties of the monster were sedulously studied by Kepler.

In a lecture to the Royal Society of Edinburgh in 1785, James Hutton, considered to be a founder of modern geology, stated, "I consider the Earth to be a superorganism and that its proper study should be by physiology." Hutton, inspired by the discovery of the circulation of the blood, noticed that similar circulation processes occur on the surface of the Earth, for example in the hydrologic cycle of flowing water—evaporation and rains.

According to Lovelock, the atmosphere is as actively regulated and organized as the skin of our bodies. Indeed, in the late 1960s, before the US *Viking* mission to Mars, Lovelock noted that life's absence could be detected on Mars simply by viewing the atmosphere. When a ray of light pierces a prism, it forms a spectrum; the composition of gases in the atmosphere of a planet will form different spectra. Lovelock held that since the gases in the Martian atmosphere as viewed spectrometrically from the Earth were conspicuously normal or in chemical equilibrium, there was no need to fly to Mars to conclude that life did not exist there. Later, after the two *Viking* spacecraft landed in June and August 1976, other scientists came to the same conclusion. The Martian probes dropped down on the soil of Mars, took pictures, picked up the soil for analysis with a special scoop, and performed a series of experiments specifically designed to find life. The sensors determined that the Martian surface didn't even contain trace quantities of the organic compounds associated with life, let alone life. At least judging from the top twenty centimeters of soil analyzed by the landers' apparatus, Mars was dead.

Meanwhile Lovelock was developing a different view of Earth life and atmosphere. Contrary to conventional geologists who viewed life as a passenger on Earth, Lovelock believes that life totally infiltrates a planet's surface: "[If] you have life on a planet, it is bound to use any of the mobile media that are available to it, like the atmosphere or the oceans, as a source of raw materials and also as a conveyor belt for getting rid of waste products and so on. And such a use of the atmosphere—and there wasn't an ocean on Mars, there's only an atmosphere—will be bound to change its chemical composition away from that of a lifeless planet."

Lovelock holds that life, a planetary phenomenon, must involve the atmosphere.

Strikingly different from Mars and Venus, both of which have atmospheric compositions that make sense to a chemist familiar with how gases react in the laboratory, the atmosphere of our Earth contains gases such as oxygen and methane that react violently together and so should vanish to be replaced by other, chemically expected compounds (such as carbon dioxide and water). Yet here on Earth, these reactive gases remain in each other's company over geological time. Oxygen, for example, can be deduced from charcoal in the fossil record. (Life leaves fossil traces in the rock record with some of the same materials—namely, chalk and charcoal—that an art student uses to draw a live figure on a piece of paper.) The charcoal shows up in the rocks regularly for at least the last three hundred million years. Now, if oxygen in the atmosphere were less than 15 percent, no match or lightning would ever start a fire. So oxygen must have been at higher concentrations than that. But if oxygen had been at higher concentrations than 25 percent, worldwide fires would burn so wildly that evidence of ancient forests would have been obliterated. Evidence of fossil forests abounds, however, so the reactive gas oxygen has remained in our atmosphere for hundreds of millions of years in the proportion of about 21 percent—even though it reacts violently with the hydrogen-rich carbon compounds characteristic of life. In a way, our planet now resembles a giant battery, kept energetic and continually recharged by the sun. Looking around the solar system, it is clearly a chemical anomaly, a space oddity.

Lovelock posits that the Earth's atmospheric composition, its mean surface temperature, the pH and salt concentration of the world's oceans—as well as many other physical factors—are under active biological control of the global biosystem itself, which he calls Gaia, after the ancient Greek goddess of the Earth. According to one line of thought, the hospitability of the environment over geological time is a coincidence; if we had not been lucky in terms of the inanimate processes stabilizing our planetary environment, we would not be around to coo over life's physiological control of the Earth's surface. But Earth without life would look as barren as the moon; clearly, life is deeply involved in geological processes in the would-be inanimate environment.

The obvious Gaia criticism is that it is difficult enough to predict an individual organism, let alone a phylogenetic colossus with a sample size of one; that the use of the term *organism* is surely metaphoric, strategic rhetoric; that life itself is not contained as the noun suggests, but is itself a process literally in evolution, incorporating new materials as it grows, colonizing rocks, producing technology, and recycling to soil and stable living as it has done, naturally, for hundreds of millions of years. The obvious criticism is that if life doesn't end at the cell, or the organism, or the ecosystem, why should it stop at the planet?

But Gaia theory in its current form does not necessarily say that the Earth is an organism. It says that global environmental variables are actively being maintained. This is what happens in many familiar organisms. Take you. Your blood chemistry is not random. Although they are made from common elements and though they are in flux, rare chemicals react to perform physiological functions within certain highly constrained boundaries. The same is true of the atmosphere. It is a sort of invisible tissue, a rare outgrowth of a complex, planet-sized system. Earth and air are not simply contiguous systems, nested nearby spheres, but integuments, active, interactive parts. They are locked into an organic dance, as linked as blood and breath, close as rain on river. The vibrant biosphere maintains its organism-like non random chemistry by energy flow.

The surface of the Earth has organism-like traits of regulation, complexity, and control, somewhat as an animal body—although differently shaped and able to walk on dry land—is chemically similar to cells, is composed of them, and feeds as they do on substances that can be used to derive energy. There is nothing mysterious about saying that Earth's surface, its biosphere, is organism-like. The atmosphere, as measurably complex as the composition of your blood or breath, reflects the chemical activity of immense crowds of growing, gas-exchanging organisms that can't help but alter the environment on a planetary scale.

Invented—along with cryogenic techniques, the microwave, and the electron capture device—by James Lovelock, the Gaia hypothesis holds that environmental variables at the surface of the Earth (such as the salinity and acid–base balance of the oceans, and the carbon dioxide and oxygen levels of the atmosphere) are controlled by the sum of planetary life. Other attributes—such as continuous removal of potentially poisonous salt from the oceans, the relative concentration of methane and other gases in the atmosphere, and even climate change—have been ascribed to the physiological nature of our planetary surface. Moreover, astronauts have reported that Earth looks alive from space. If it looks like an organism, reacts to perturbations like an organism, and slowly grows under the sun, perhaps it is an organism—although of course not an animal in any familiar sense. One of the arguments against Gaia's status as a true organism is that organisms reproduce. But this is not always true. Mules, the offspring of donkeys and horses, don't reproduce. More to the point, the Gaian macroorganism appears to be capable of reproduction.

This might be just another cockamamie new age idea were there not rich evidence for it. But how does a seemingly stupid and inert planet keep itself cool in the face of a brightening sun? How does it work to remove salt from its roiling blue oceans? How does it manage to maintain, despite increasing pollution, the reactive atmosphere that provides the chemical fodder for fireworks, lightning bolts, rock concerts, and your comprehension of this sentence? But then we are largely ignorant of the fantastic organized mess, the cellular commutes and migrations, the inner governments and alliances of our own bodies—this intricate physico-chemical marsh from which a sense of "me" seems to arise. The anomalous atmosphere is an extension of the organization of the complex energy-using forms on the planetary surface. If intelligent aliens existed (and could be bothered), perhaps they would take us adults aside as children are sometimes taken aside, and give us a new facts-of-life lecture. The time has come for us to know that we are in it deep. Profound is the chemical organization of Earth's surface, which is more reminiscent of a beehive or body than a spherical rock with some life on it. On the other hand, perhaps there is nothing to say to us overweening dweebs: Aliens, if they exist, wanting to be taken to the leader, might communicate directly with the Earth, four and a half billion years our senior. Perhaps they already have.

Stanislaw Lem (1921–2006)—a Polish science-fiction writer who wrote in Russian and was thought by science-fiction writer Philip Kindred Dick to be an invention of the KGB—wrote a novel called *Solaris*. Twice made into a movie, *Solaris* centers on a living ocean able to transmit messages—in effect embody an imagined lost family and lovers—to an orbiting spacecraft. In *Solaris* the ocean of a distant planet detects humans nearby in its orbit. Willing to make overtures and to listen to its visitors in the idiom of their own minds, the distant and liquid world sends living copies of people who are no longer there onto the space station of scientists. This of course confuses them, as the physical embodiments of the missing or departed people, taken from the scientists' memories, are not to be taken at face value. They fulfill a wish for companionship, and the depth of previous affections is compromised by the ersatz status of the interlopers as mere human greeting cards.

Solaris, the mythical living ocean imagined by Stanislaw Lem, may not in the final analysis be more bizarre than Gaia, the real living planet detected by Lovelock. As Earth and science evolve, Solaris looks less like an old black-and-white Russian science-fiction film or the tale of a remote and planet-size alien being, and more like a coming attraction.

Truth is stranger than science fiction.

Earth is a complex
thermodynamic phenomenon with clear physiological traits. If, with Maurizio Ferraris, we cannot even say that we are alive, then we cannot say that the Earth is alive. But if we are, then so may it be. And if it is alive, it may be intelligent. Is the entire universe, in fact, alive? Is it intelligent? Has it always been?

I am no scientist and so everything I say should be taken with a grain of salt—salt such as that which, running along rivers, collects in the oceans and should thus increase in concentration to the point at which it is poisonous to many familiar organisms, but is somehow mysteriously removed, probably by the organisms themselves. The entire biosphere seems to be a complex thermodynamic system—"eating" sunlight, cycling materials, and growing as it produces waste, mostly as heat, into the cosmic environment.

What we call life is from a certain perspective a very interesting kind of complex thermodynamic system. Thermodynamics is the study of how systems handle energy. Most systems simply seem to come to a steady state of relative equilibrium. This is what the second law of thermodynamics is generally assumed to say: that complex organizations run down, their capacity to generate work inevitably decreasing as they become more disordered, boring, and uniform.

But a class of systems—real complex systems—naturally grow and organize their local surroundings. To this class, which Eric D. Schneider and I wrote about in *Into the Cool,* belong Bénard cells, Taylor vortices, Liesegang rings, Belousov-Zhabotinsky reactions, salt fingers in the ocean, Hadley cells in the atmosphere, and tornadoes. The function of all members of this somewhat eclectic group—and note that none of the entries in the above list is, technically, alive—is to maintain and grow using available energy, which measurably helps bring complex environments to equilibrium, producing atomic chaos, or entropy, in accord with thermodynamics' famous second law. As Gifford Lecturer Sir Arthur Eddington famously said in 1927:

> Entropy continually increases. We can, by isolating parts of the world and postulating rather idealised conditions in our problems, arrest the increase, but we cannot turn it into a decrease. That would involve something much worse than a violation of an ordinary law of Nature, namely, an improbable coincidence. The law that entropy always increases—the second law of thermodynamics—holds, I think, the supreme position among the laws of Nature. If someone points out to you that your pet theory of the universe is in disagreement with Maxwell's equations—then so much the worse for Maxwell's equations. If it is found to be contradicted by observation—well, these experimentalists do bungle things sometimes. But if your theory is found to be against the second law of thermodynamics I can give you no hope; there is nothing for it but to collapse in deepest humiliation.

Life may be a sort of glorified heat producer, keeping itself cool locally and producing more entropy than would be the case without it. Whether life has an ultimate spiritual purpose is not clear, but scientifically life clearly belongs to this group of energy-using, material-cycling systems, and thus seems more than ever a natural physical process—whether considered at the level of the cell or organism or planet as a whole. Life differs from nonliving systems such as whirlpools and intricate chemical reactions that cycle matter and grow more organized as they produce entropy—basically atomic chaos as heat—into their surroundings. Although within the same class, living matter has the genetic capacity to repeat its structure and thus prolong its efficient entropy production—which, as Sir Arthur suggests, is favored by the universe. So unlike what you might have been told, there is no contradiction between life and the second law. Life is organized, yes, but part and parcel of that organization is its production of entropy, which shows up in the biosphere's ability to turn solar energy into cell structure, produce heat, and, at least when healthy, keep itself cool.

Life on Earth, whether or not you consider it alive and call it Gaia, is perhaps the most stunning example of such a system. The material structures of life are repeated, and the system as a whole grows in spatial extent, complexity, and, yes, intelligence. Intelligence, conscious or unconscious, human or non-, allows us to tap into and exploit energy sources, growing locally as we spread atomic discord into our surround.

P. K. Dick was so intrigued with the notion of entropy—a physical measure associated with the tendency of energy to spread out over time—that he invented words like *gubble* and *kipple* to commemorate matter's tendency to replace organized formations with formlessness. As a single system that may or may not be alive, the entire surface of the Earth is remarkably organized from a chemical or thermodynamic standpoint. But this organization, like that of individual organisms, is natural for complex systems that cycle materials as they "feed" on energy flow. Biospheric complexity generates extraterrestrial "kipple" in the form of heat. Whether considered an organism or not, the biosphere's striking physiochemical organization appears a natural process involving the cycling of matter as energy flows. Accompanied by electric guitar, Jimi Hendrix sang, "There ain't no life nowhere." Ultimately the category of life may obscure our view of natural processes in their cosmic setting. "Life" is integrally and eternally connected to the "nonliving" universe from whence it derives. Earth is no more a rock with some life on it than you are a skeleton infested with cells.

The Earth may be alive
just as the universe itself is in science-fiction writer Philip K. Dick's notion of VALIS, an acronym for Vast Active Living Intelligent System. In the speculations that follow I operate on the precept, formulated by geneticist J. B. S. Haldane that, "The world is not only queerer than we suppose, it is queerer than we can suppose." No matter how strange or wonderful our imaginings, they may fail to capture the full strangeness of existence. The cosmos, gargantuan as it is, may yet be part of a larger whole we cannot perceive. Indeed, this whole may be infinite.

My relative pettiness as a human being has not deterred but spurred me in my quest, as a former amateur magician, to whisper to reality in an effort to get her to reveal her secrets. If I have erred, I plead humanness, because to err is human—and besides, to be located as a discrete being, a human in a specific place and time with a particular perspective, is necessarily errant. The whole may remain forever occluded from us due to our status as mere parts. But being parts means we can err. We can wander, rove, move in space and time. Who knows whether this wandering and roving is not ultimately illusory. Without it, however, there would be no search, no adventure. The word *planet* comes from the Greek for "wanderer"—because the ancients noticed that the paths of the planets varied relative to the "fixed" stars. Earth, too, wanders.

Truth, as I have been saying, is not only stranger than fiction; it is as strange as the most radical science fiction. We criticize people when they say they think they are the center of the universe. But each of us feels that way. God's center is everywhere but his periphery is nowhere. So wrote German mystic Meister Eckhart. Each of us seems to be the center. We see things from our personal

The universe curiously seems like an almost complete puzzle with only a single piece missing. Each of us, like a man miscounting party guests because he neglects to count himself, or a woman searching for keys already in her pocket, seeks the answer. What you may forget is that it is right in front of you. In fact, it is you. *You* are the missing puzzle piece.

point of view. We are dejected when bad things happen to us. But perhaps the wisest advice there is (except for the time-honored "A word of advice: Don't give it"), and the most scientific, is "Don't take life personally." Each of us tends to take things personally. We may empathize with the viewpoint of others in love and fiction, but we inevitably sense the world from our own vantage point. Perhaps this is the only way perception can happen. Only saints have total empathy and no one identifies with saints because they are, well, so impersonal.

From a mystical as well as logical point of view, it may be that it is impossible for the whole to observe itself. A universe not divided into parts has no chance of self-observation. It never knows that it is. Thus illusion—the illusion of separation of what is one from itself—is constitutive. The universe can only know itself if it is split from itself. An unsplit universe can never know of its own existence. Here is the dual secret of love, which brings us together, and awareness, which requires that we be separate from what we observe. Our lives here on Earth exist in a state of tension between the two—we hanker after what we love, but union with it would lead to our dissolution. Knowing this, we maintain a certain distance, at least until the time of our death, which deprives us not only of our life but of the constitutive illusions that govern it as well. A universe that can only know itself if it is split from itself: Surely, you may say, this is pure metaphysics. But there is backing for it from Einstein's still-living colleague, the prominent American physicist John Archibald Wheeler (b. 1911). Wheeler has a diagram that exemplifies the human condition. It is an abstract, vaguely Dali-esque horseshoe-shaped thing looking like a crowbar with an eyeball peering back on the U-shape to which it is connected. We are that eye. The figure connected to an eyeball gazing upon its extended self is none other than the observer querying his impression of separateness. Conscious worlds may be like that. Observation in general may be like that. Only a universe that can be separated from itself can perceive itself. Perception requires a subject and an object, even if, as proponents of Eastern religion like to say, "I am that." Yes, you are. Deception—the convincing illusion of the separation of the individual from the environment that he observes—is the precondition for perception.

Only the part—whose separation from the whole is ultimately illusory, because it must, in the end, in space and time, be connected to that whole—can perceive the whole. We are part of that whole, as is the evolving biosphere. We are tribal nomads dwelling in the Holocene, that geological snapshot starting only eleven thousand years ago, during which time our brand of hominid has spread like wildfire, diverting other species' resources and Earth's energy to fund our own rampant reproduction.

The world

is a strange and beautiful place. And magical. The same reality admits of multiple perspectives. We can see it from the front of the stage, like spectators at a magic show. Or from the back, like scientists figuring it out. But science learning the secret—such as the strange correspondence between mathematics and phenomena—may lead to further mysteries. "The most beautiful experience we can have is the mysterious," wrote Albert Einstein, perhaps the greatest scientist of the twentieth century. "It is the fundamental emotion which stands at the cradle of true art and true science." "Perhaps the only real thing about him was his innate conviction that everything that had been created in the world of art, science or sentiment, was only a more or less clever trick," wrote Vladimir Nabokov, perhaps the greatest writer of the twentieth century. Illusion is the first of all pleasures, wrote Oscar Wilde, perhaps the greatest wit in the history of the English language. But as quantum physicist David Bohm says, science is about finding the truth *whether we like it or not*. Sometimes we don't want to know. The whittling down of something wondrous

Hol·o·cene *adj.* Of or belonging to the geologic time, rock series, or sedimentary deposits of the more recent of the two epochs of the Quaternary Period, beginning at the end of the last Ice Age about 11,000 years ago and characterized by the development of human civilizations.

to something mechanical can be disappointing. The ultimate example of potential disillusionment leading to depression is the knowledge that we may die. A lot of people don't like science because it doesn't tell them what they want to hear. Like a new bride, they want to hear that they are special, but science says no, we're not. The situation reminds me of a Groucho Marx joke from the time when he was a narrator on the quiz show *You Bet Your Life.* A husband came back to see how his wife answered the question, *From which direction does the sun rise?* She said west and of course her husband was chagrined.

She saw his expression and heard his disparaging, "The sun always rises in the east."

"Not in our neighborhood it doesn't, honey."

Like the disappointed newlywed, we are chagrined to confront the scientific litany of our cosmic ordinariness. The Earth is not the center of the universe. We share 99 percent of our genes with chimps. The atoms of our bodies are mostly hydrogen—the most common atom in the cosmos. The other main atoms, carbon, oxygen, nitrogen, sulfur, and phosphorus, are also common. The element used to treat manic depression, lithium, is the third lightest atom and one of the first to appear, according to cosmologists, after the Big Bang. Perhaps God knew manic depression was coming and wanted to get a head start preparing the medicaments. We did not fall from the sky but came out of the cosmos with its mélanges of matter; we are not pure but latter-day taints, the progeny of microbial miscegenation and skulking apes. And as suggested above, and detailed below, even the energetic processes in which we are involved are not original. Our energy-garnering ways are shared with other matter-cycling manifestations of the second law; not only are we organic dreck, cosmic debris as Frank Zappa once put it, but we are, in our swirling operations, our backing and filling to acquire the resources necessary to preserve our system of cyclical resource acquisition, typical. We may just be another natural form of growing thermodynamic system. Not just the stuff of our bodies, in other words, but the processes in which we are involved may be frightfully common.

The temptation to consider ourselves special is overwhelming.

Even my father, in his Gifford Lectures, after titling one "The Retreat from Copernicus: A Modern Loss of Nerve," retreats into the sirenlike sanctuary of human specialness, assuring his readers (his listeners) that humans are the dominant species. With which most people of course would be as quick to agree as would the ladies' man to tell his lover *de nuit* that she is the most beautiful woman in the world. We may not be special but, well, we are special. The temptation is tenacious, like the spiral eyes of some secret Narcissus beckoning an adolescent to gaze again into the bathroom mirror.

Yet as that same adolescent we all were may have heard, "Consider the source": When the species in question is judging itself, despite all the careful scientific qualifications, to be the latest, the greatest, the most evolutionarily significant thing since sliced bread, we need to take a step back. Is it possible, perhaps, that we are not the dominant species?

In fact the focused ways in which we seek food, shelter, mates, money, and safety for our children are all ways to maintain our particular form of cyclical material organization. We may well be, despite the creative audacity of our delusions of grandeur, just one more nonequilibrium system that produces more entropy than would be the case were it not to exist—were it to be replaced by a mere random collection of particles. In a quote sometimes credited to Mae West, "Life is not one thing after another; it's the same damn thing, over and over again." Our cycling of matter to preserve our particular form, I would wager, partakes of a universal process. Call it the dilemma of the loose species. We would like to think we are virginal and pure—a level below the angels in the great video game of traditional religious cosmology. But the reality is more like the Madonna song: We are *like* a virgin. This isn't that bad because it is ultimately more fun—and, I would say, more religious—to own up to our cosmic promiscuity. *Religion* comes from the Latin words for "relinking." Noticing the extent to which we are part of, rather than apart from, the cosmos gives us new fortitude to withstand those well-noted feelings of alienation, the existential malaise that the German philosopher Martin Heidegger called "thrownness."

In other words, not only the stuff of life but also its process—

including its purposeful behavior to seek food and energy sources, and our own tendencies to do so—is not special. The world isn't all it's cracked up to be. We did not come out of the cosmos like a virgin birth. We are different in particulars but general in type. Despite the widespread acceptance of evolution among educated people, there is a sort of atavistic cultural mechanism by which we try to insist that we are in some way special. If not our spirits then our opposable thumbs, our ability to speak and write, our culture, our technology, our brains or brain-to-body ratio, whales excepted. What is amusing about this ad-libbing, this hemming and hawing to find something, anything, special about *us,* is its desperation. We protest too much. Indeed, the need to separate ourselves in some final way from natural phenomena runs counter to the spirit of evolution itself: Charles Darwin's connection of humans to other life-forms through common descent is a movement of connection, just as Copernicus's placement of the sun at the solar system's center, far from sending us to our room, crying about how the universe doesn't like us, should make us jump for joy that we truly belong to the biggest, greatest club there is. Woody Allen said he always put his wife under a pedestal. We do not need to be manic depressive. We are neither gods nor beasts but tribal nomads on an increasingly crowded planet that remains, nonetheless, full of potential.

The loss of our theological virginity, our would-be divinity, is an illusory catastrophe. We never were virgins. As that brilliant American wit and daughter of a southern senator Tallulah Bankhead* put it: "I am as pure as the driven slush." Let us take the epigram of the Alabama socialite to heart. We, too, are as pure as driven slush.

And this, I submit, is a good thing.

Now, I'm sure

some readers will want to throw this thankfully flimsy book across the room when I say this, but as the great Danish quantum physicist

*Ever amusing, Bankhead also said, "My father warned me about men and alcohol. He never mentioned anything about women and cocaine."

Niels Bohr said, the opposite of a great truth is another great truth. I've always held that people's greatest strengths are their greatest weaknesses. I, for instance, have a weakness for epigrams—which, while fun, may derail the logical thought necessary to figure out, or at least approach, the secrets of existence. Nonetheless, we've come this far and there's not all that much farther to go. Once we've gone through the scientific wringer and gotten over ourselves, realized that we are not so very special, then perhaps it is time to come back and take a fresh look. I do not necessarily consider what I'm about to say contradictory in an absolute sense. I would more likely describe it as temporarily contradictory. In linear time, it contradicts itself. But when all is said and done, it doesn't.

So we may be able to dialectically return to our original impulse to consider ourselves central and find that it has merit. Science so often says we are ordinary, but Protagoras says man is the measure of all things. Strangely, both are right. There is a relativity of truth here. Let me give you an example. Let's return to the quote, "God's center is everywhere; his periphery nowhere," attributed not only to Eckhart but also, in various forms, to Empedocles, Augustine, Cusanas, Bruno, Timaeus of Locris, and Alain de Lille, a twelfth-century theologian.

Something similar occurs with the idea of Earth being the center of the universe. According to modern heliocentric theory, the planets orbit the sun. According to what we learn in school, our ancestors were wrong. They thought the sun orbited Earth. Yet as it turns out, they were also right.

How can they be both right and wrong?

Well, relativity theory tells us there is no privileged frame of reference for motion. When Einstein sat in the famous train in Bern, Switzerland, and saw another train move and mistook it for his own train moving, he realized that it was more than a misapprehension. Objects do not move absolutely but relative to each other. We are

This sentence has two mistakes. The first mistake is that it says that it has two mistakes, when in fact it has none. The second mistake is that although it has only this one mistake, it says it has two. This makes it true after all.

thus within our rights to regard Earth as central, according to modern physics.

Apparently we are special after all.

I thought this idea was original until I talked to my mathematician-poet friend Steven Shavel, who thought it had been original with him. Then I read it in *Physics and Philosophy,* a 1948 book by astrophysicist Sir James Jeans. Einstein's thought process was amazing. He came to the idea that space is curved by thinking of what happens to parachutists falling parallel to each other from very high-flying airplanes. Although the parachutists fall straight when they jump, they nonetheless converge upon the Earth. Mass is a curvature of space. Earth may be regarded as the center of the solar system.

Einstein may have been partly inspired in his thoughts on relativity because of his personal experiences in Austria outside the social elite: Were those who thought, believed, and behaved special *really* special? And yet Einstein, it seems, *is* special. Paraphrasing Shakespeare, we can say that science has shown that there is more in Heaven and Earth than dreamt of in our medieval philosophy of being specially created at the center of the universe. On the other hand, as the mystics have long argued, and relativity theory supports, the universe's center may be taken to be at any point. We certainly feel central. Thus, connected to a possibly infinite universe, we may be central—but no more central than anything else in the connected universe.

The more things change, the more they stay the same.

I am reminded of the well-known card magician and creator of optical illusions Jerry Andrus. Known for performing waterfalls and flourishes with phosphorescent playing cards under black light at his home—the same place he has lived for several decades—Andrus told me a story. He said that when he was a kid they told him that the country he was living in, the United States of America, was the greatest in the world. Later, in elementary school, he learned that his state, Oregon, was the greatest in the country. Then, in high school, what with the football teams and cheerleaders and all, he heard that his town, Albany, was the greatest in the state—better than Eugene, better than Portland even. It was enough to make a guy—or at least the curious, creative, illusion-generating mind of Jerry Andrus—suspicious.

I capitalize

Earth to conform with the NASA designation, which uses the word to refer more to our planetary blue home than, as of course is also valid, a handful of dirt; Earth. In Gabriel Garcia Marquez's *100 Years of Solitude,* a pregnant woman consumes it. The practice, pica, is most common in Africa and the southern United States. The Pomo of Northern California mix earth with ground acorn. Birds sip at natural reserves of clay; the bitumite collected and sold in natural food stores is a powerful detoxifying agent able to kill microorganisms and leach many times its weight in heavy metals. Tilled by insects and earthworms, giving rise to orchids and roses, the soil is so full of life, it *is* life—more like a planetary skin full of cells and circulating nutrients than mere dirt, however much even (or especially) the pregnant may crave it for its calcium, phosphorus, potassium, magnesium, copper, zinc, manganese, and iron—elements required to produce new life. From Earth's wellspring come DNA, RNA proteins, the ten thousand things and thirty million species, from it uncounted awarenesses, traces upon traces, the concerted actions of a well-connected but unconcerned cellular citizenry.

Earliest life may have come from the sticky sides of submarine smokers closer to the surface in Earth's earlier, shallower oceans. Feeding on natural iron and sulfur reactions, surrounded by a primordial stink (the result of hydrogen and sulfur compounds) more appropriate to a poorly ventilated latrine than the golden cradle of the chemical process that would ultimately give rise to humanity, early life-forms spread. Made of Earth, life carried on inside itself the chemical reactions that were, evolutionarily speaking, taken from the external environment. Using energy from the sun to maintain the complex process of maintaining and making new bodies, life spread from its possible lowly beginnings as chemically active crud.

The heat vent origin-of-life story may seem more like science fiction than the dual versions of the Good Book—and it is, God knows, only one of many evolutionary scenarios. However, it has going for it the fact that inanimate iron-sulfur chemistry is not only evocative of, but also seen in, all modern living cells. Moreover, the thermodynamic process of growth using available chemical energy is a viable missing link. Also, but not finally, the most primitive kinds of modern bacteria, as judged by their RNA, are found in extreme environments such as boiling springs as well as in places without any atmospheric oxygen. They may thus be the long-lost remnants of cell ancestors adapted to a hotter, more violent Earth without free oxygen in the atmosphere.

"Angels and demons and sloppily splitting precursors, oh my!" The picture of our honorable seniors, not necessarily even fully living yet, adhering to the sides of sulfur-steeped substrate in a volcanic milieu replete with red lava is one more evocative of Hell than Heaven or the Garden of Eden. Opposites attract. Such are the vicissitudes, the prejudices, the easy opposites of the creatively flawed and dichotomous human mind.

Earth is what we make of it. I live on a planet inside my head on the planet inside my head. And so on. Earth forms and re-forms. In and out of our imaginations. All of humanity may be a specialized tissue, a disperse Cycloptic organ through which Gaia narcissistically checks herself out in the depths of space. We are simply along for the ride.

The universe, scientists reckon, is some thirteen billion years old. The Hadean ("Hellish") eon saw the beginnings of our planet in a gravitational swarm of swirling lava, volcanic violence, and meteoritic bombardment. The Archean (from the Greek for "ancient," as in *archaic* and *archeology*) saw the rise of life, as attested to by isotopic evidence and, more directly, microfossils of those poor excuses for a deity and awkward ancestors, the bacteria. Due to their infectious swapping of genes (appropriated in very small part by modern biotech firms), however, as well as their hardiness, rapid growth, and extreme metabolic diversity, these organisms took over the planet, symbiotically merging to make protoctists—the cells that would, evolving in separate lineages, go on to become animals, fungi, and plants.

Two billion years ago there may not have been life on Mars, but there certainly was on Earth. If you watch a certain DVD of early life on Earth, you can see the early nucleated mastigotes going at it, not fighting but trying to swallow each other. Microbes, they have no immune systems, and their wallowing is as pleasurable as it is unsatisfying. Merging membranes, they double the chromosomal numbers of their nuclei. Now they are doubled cellular beasts, pulsing in the primeval slime beneath a sky that smells like an industrial cesspool. The sun seems small and the planet strangely tropical, inhabited only by a voluptuous scum. No plants or animals could survive the strangulating lack of oxygen in the atmosphere. But the hypermastigote-like beings thrive. They throb in each other's membranes, two in one. And unlike other beings who have similarly cannibalistically merged, this couple is lucky. When they divide, half the chromosomes go to one body, the other half to the other. If we were flip, we might call one half Adam, the other Eve. Or, to quote Bette Davis, "with separate bedrooms and separate bathrooms, I give them a fighting chance—[cackle]."

What was this? The invention of reproductive sex.

Sex of the sort that unites sperm and egg and is required for reproduction probably began sometime in the Proterozoic eon. By this time the bacteria had grown the surface of the planet to create an Earth system that, though nothing other than organisms in their environment, displays global physiological features such as temperature regulation and control of the atmosphere's composition. It was the noticing of concerted physical action not reducible to standard chemistry and physics at the global level that prompted James Lovelock to develop the Gaia hypothesis. Considered as a global physiological entity displaying more than the sum of its organismic parts, Gaia, as the largest living being in the solar system, probably developed sometime during the Proterozoic eon.

Fungi are part of the story, too. Fungi grow on Earth, they recycle dead (and sometimes still living) bodies back to earth, and, quite literally, they are parts of the earth. As the great mycologist Paul Stamets points out, animals and fungi only separated from common cellular stock 530-odd-million years ago. Fungi digest their food on the outside, breaking it down with enzymes before taking it into their growing thread bodies. The world is their stomach. They perhaps partnered with green algae (not true plants, because they don't form embryos) to form the first land vegetation. *Pilobolus* spores grow on cow waste, long-jumping from cow pat to green grass in a never-ending cycle that requires traversal of the bovine intestinal tract and a delicate balancing act between being eaten and being devoured. Other difficult-to-digest fungi, psilocybin mushrooms, growing in similar pastures, reliably produce hallucinatory, even quasi-religious, visions in their human ingestors. The genus *Cordyceps*, a parasitic fungus that forms clublike appendages, is known for discretely attacking and chemically disabling specific woodland species, from insects such as moths and butterflies to other fungi such as puffballs. The infamous *Cordyceps lloydii* secretes a chemical into the ants it infects that leads the latter, as if under the influence of some science-fiction mind control mechanism, to climb to the top of the rain forest canopy where they attach themselves to leaves and die, prior to the fungus's fruiting body, its mushroom tip, sprouting through the top of their heads; wind distributes the mushroom spores to new ants, repeating the cycle. *Cordyceps sinensis,* sold by Stamets's MycoMedicinals, was observed by Tibetan herders 1,500 years ago conferring energy and "passion" on the yaks that would seek out and eat the mushrooms. Some ants and termites have evolved subtle means of detecting *Cordyceps* infestation. Queen-guarding soldiers in these social insect species will remove infected members of the colony, pushing them off branches or otherwise eliminating them to protect the group. Amateur mycologist and English teacher Michael Kuo points out that such sophisticated behavior compares favorably to the extraterrestrially infected crew in the movie *Alien*: If Sigourney Weaver and the others had sensed that John Hurt was taken over by an alien parasite, and killed him and quarantined his body, they would have eliminated their chances of contagion and saved themselves lunch money and the informational expense of an interview with a hidden alien holed up in a human body. Once again, one does not have to search far to find examples from the nature beneath our feet that trump the lofty flights of the science-fiction imagination. Fascinatingly, the Chinese National Track and Field Team suggest that their ingestion of *Cordyceps sinensis* (the kind liked by yaks) is at least partially responsible for their 1993 breaking of three World Records.

The coevolution between fungi and animals is extremely deep, and fungal mycelia and their chemical byproducts may even unwittingly influence some of our most seemingly human and animal behaviors and desires. Cats cover their feces; humans bury their dead: are we unconsciously fulfilling the needs of the fungi, who can't wait to get their mycelial mitts on our recyclable remains? I once wrote a poem—and still think it would make a cool science-fiction story—about a master plot by fungi, secretly intelligent despite their literal low profile, to use humans to conquer the stars. Or at least colonize the other planets: What wouldn't be better for this mysterious terrestrial life-form than to get its paws on the surfaces of other alien orbs—to begin chewing up the Martian regolith, say, or begin spreading their subterranean tendrils laterally across the rich organic substrates of primate-imperiling Jovian moons? My one encounter with a UFO turned out to be the reddish light atop a ski lodge in Telluride, Colorado. Perhaps the hero, with a taste for fine wines and cheeses, and fungally fermented champagne, suffering from a weird genetic hybrid, would find out that the uncanny aliens were in fact of local origin.

Profane yet sacred, dirt is clean, sweet smelling, and fertile because it is chemically treated by all-natural bacteria and recycling fungi. Only our historical hankering for civilization, our growing population-exacerbated experience with the relatively rare number of disease organisms, makes us think Earth is ours. Whatever we do while we're here, we belong to the Earth, and will return to it when we are gone.

Not surprisingly, the story of life as understood by modern science reads like science fiction. But then, what is the cultural alternative? Terry Carr, an editor at the pulp science fiction publishing house Ace Books, used to joke that if the Bible were submitted as a manuscript, he would have to, for marketing reasons, instruct its author to cut it to twenty thousand words and give it a more catchy title. The Old Testament would no doubt reach a more receptive audience if it were called *The Master of Chaos*. Instead of the New Testament, he would probably go with *The Thing with Three Souls*.

The Proterozoic eon—from the Greek for "pre-animal"—saw the

evolution of trilobites, insects, plants, and other relatively large forms of life. The Phanerozoic (from the Greek for "visible," as in the word *phenomenon*) eon is the one we're living in now. Just as we can be fooled in magic by not seeing what is going on behind the scenes, or under the table, so we are missing the context of our situation here on Earth if we don't take account of time. You can't expect to walk half an hour late into an intricate thriller and understand it, or divine the plot of a mystery novel from the last twenty pages. So, too, the story of life. Not only is human history insufficient to give us a coherent view of our place in space and time, but even animal history may be insufficient. To have a better chance of understanding who we are, where we are, why we are, what we're doing, and where we're going, we have to step outside the box. We have to take a peek from backstage. And backstage in this case is a less limited view of time.

Even those who are taken with the theory of evolution and the need to involve it in deep explanations of human affairs tend to restrict their vision to at best half a billion years ago, the Phanerozoic eon, which includes the Paleozoic era with its Cambrian, Devonian, and Carboniferous periods (among others), the Mesozoic era with its Cretaceous and Triassic-Jurassic mass extinctions, the Cenozoic era with its Pliocene, Miocene, Oligocene, Eocene, and Paleocene epochs. The evolutionary psychologists with their emphasis on the prehistoric background to modern human thought patterns confine their speculations to our formative years in the Pleistocene epoch, 1.8 million to 10,000 years ago. Through a kind of logarithmic self-centeredness, we home in on the present as if the entire universe devolves from it, rather than the other way around. In a way—because we are here and now, and the great swatches of time remain outside our purview—we can't help it.

How little we know how little we know: Despite their arguments, science, religion, and philosophy tend to agree on one thing: the wisdom of humility. For science such wisdom is embedded in its successful methodology, which is to say, in principle at least, the jettisoning of hypotheses that don't match up with the evidence. For religion the humility of wisdom is formalized in the church fathers pronouncing pride to be the mother of all sins or, from an Eastern perspective, the focus in Zen on "no-mind" and "nothing." In philosophy the maieutic or Socratic method involves trying to find out, learning about oneself, and knowing that we don't know. It is as if humans collectively were afflicted with the intellectual equivalent of myopia. We have knowopia—the more we know, the more we know that we don't know.

The difference between a scientist and a philosopher, it is said, is that a scientist is someone who learns more and more about less and less until he knows everything about nothing, while a philosopher is someone who learns less and less about more and more until he knows nothing about everything.

I have tried to take a middle road. . . .

These then will have been my notes from the Holocene, the epoch that began some ten thousand years ago, roughly with the dawn of agriculture and the weedy growth of the human, which Montaigne called a "thinking reed," across the face of this fascinating planet.

WATER

Do not think of the water failing; for this water has no end.

—RUMI

Invisible or translucent,
refracting light in a rainbow or fluffy white, it is the key ingredient in tears, composing, as Ani DiFranco sings, "78 percent" by weight of our bodies, "even our pumping hearts." Although plentiful on Earth and found elsewhere such as in large ice deposits at the north and south poles of Mars, reports are that it has also been found outside our solar system, on the Jupiter-like Planet HD 209458b, one of some two hundred extraterrestrial planets so far discovered. From space, Earth's surface can be seen to be dominated by it. No life here can remain active without it. The main ingredient, the heart of flesh and mucus and, indeed, of cells, it is the element of dissolution, the medium of transmission, the reason for bodily fluids. Life and water are so close that life basically is a form of water, although a unique form with some fascinating impurities. Terpenes from trees, the main ingredient in essential oils, as well as bacteria that grow on the surfaces of leaves, can lift off and serve as the basis for formation of water droplets, precipitation, as cloud condensation nuclei—until rain falls, completing one of the many would-be strictly geological cycles in which life no longer so secretly operates.

Water, H_2O, the hydrogen taken into early life-forms and the reactive oxygen let loose after the metabolic discovery. Earth's atmosphere, low in carbon dioxide and high in hydrogen, is itself an offshoot of the water-using, oxygen-releasing trick life played, and keeps playing under the name of photosynthesis, on water: using it to make more of itself. And so the chemical incest continues. Water, the medium in which life evolved, but whose relationship with life became more intense—some eighteen hundred million years ago, according to the fossil record—when life, for the first time, found it could use the hydrogen of this medium to make more of itself.

Water, the clear medium we see as our selves in the blue planetary mirror of the Earth from space.

Water—the reflective medium of our eyes.

Life is a form of water that has taken to the shore and woods, the savanna and hills, the rain jungles so rich and wet with life. Watery life's ancient ocean expansion has led to an occupation of the land. Even in our suburbs with gardens and houseplants, or in the midst of the grittiest city, there lie the waterworks, the plumbing and drainage systems, everything and the kitchen sink, down to the tubs and toilets that Milan Kundera called the swollen ends of sewer pipes.

Life had been living in water all along, without realizing its extreme potential: the discovery of water's ability to serve as a hydrogen source for life's supersubtle carbon-hydrogen bodies. From bacteria and their metabolic tricks to the sexually reproducing "higher" forms, watery life has had its way with Earth.

Now the two are one.

But even on its own, water is a most prodigious player. Indeed, we flatter ourselves to consider water a form of life. Life is more a form of water: In addition to the interconnections of the two "substances," water's chemical formula can be precisely specified. Life, however, despite its known genetic composition and bodies made from characteristic proteins and enzymes, hardly varies enough from water and varies too much in itself, to be considered from an objective material standpoint strictly its "own" substance.

This should not surprise us, considering our earlier unsettling discoveries as to life's cosmic commoness.

As a thing, life is not that different from water. But life is not a thing; it is a process in which different materials are superadded, making life more than it was—more intricate, more encompassing, more energetic, and more technological. Among the elements that life—even before technological human beings—integrated into its expanding reproducing bodies are calcium, strontium, phosphorus minerals, iron ores, and possibly gold, everything from your pearly whites to the magnetite teeth of chitons.

Chemically cycling energy-dependent life is a curious phenomenon. But even were life not on this ocean world, this planet would react in strange ways due to its wet nature. Water, for example, freezes, and thus reflects more light and heat to space. Although giant mirrors carted to space have been proposed as one stopgap solution to the greenhouse effect, such reflectivity—when derived from ice—could theoretically plunge Earth into an ice age. Theoretically, if enough ice freezes, reflecting sunlight and cooling a watery orb, a feedback cycle would arise that could collapse Earth into an ice ball. Our real water world doesn't work that way, not only because this is a geological oversimplification but also because life exerts a greater impact on the Earth's surface than is recognized by mainstream geology. Oceans themselves fill space made possible by plate tectonics, itself likely life-dependent because calcium carbonate—made of dying ocean microorganisms—lubricates the continental plates.

As Lovelock points out, life is more abundant in the northern and temperate regions of the ocean than (as you might think) at the equator. Life, literally, is cool. That's the way it likes it—which makes sense when you consider life a thermodynamic phenomenon whose main job is to produce heat as entropy. Although there is more terrestrial life in the tropics than in the northern latitudes, such life is concentrated in rain forests whose net effect seems to be to cool the planet by forming white, reflective clouds above them.

Before you say I'm all wet, let's take a look at the devious feedback that oceans display on their own. (Parenthetically, we shouldn't be surprised that life is a cyclical chemical process in an astronomic environment dominated by cycles, from the esoteric possible influences of the galaxy and stars to the more down-to-Earth light and energy cycles of the shining sun and phase-making moon. Cyclical life, once we lose our bias, may simply be an outgrowth of a cyclical universe.) Ice cooling the planet leading to more ice and more cooling is a positive feedback. So is ice melting and, with a lower albedo or reflectivity, leading to warmer temperatures and more ice melting. But there are negative feedbacks, too: warming leading to cooling, and cooling leading to warming. Because life lives only in a certain temperature range, it is interested, as it were, in being neither too hot nor too cool. This is true of us both consciously and unconsciously. Consciously if we are too hot we will turn on an air conditioner or fan, or take off a sweater; if too cold we turn up the heat, or move to a hotter clime, or buy a new jacket and mittens. Unconsciously if we are too hot we sweat. And of course if we are too cold, we shiver. Also, if we're too cold, our mammal bodies will direct the blood to the vital organs in an act of rationing. This leads to frostbite. Although we, as warm-blooded animals, are more sensitive to temperature changes than some other organisms such as reptiles, all organisms must navigate between the Scylla of overheating and the Charybdis of freezing. Life only functions as life within certain temperatures, although at extreme cold temperatures organisms may preserve their energy-dependent process, and come to life if and when temperature, water availability, and other conditions become favorable again.

The "inanimate" Earth often works likewise. Global warming and cooling are water-mediated. If enough polar ice melts, for example, because of heating, the Gulf Stream, which brings warm water to England from the Gulf of Mexico, could be shut down, leading to cooling. Even without the intelligent-seeming, physiological feedback systems of the body and consciousness—for example, putting on your jacket when cold or sweating when hot—the complex Earth system appears to have ways by which it may, however partially, thermoregulate. Even the planetary surface considered as an inanimate system, in other words, may have the capacity for negative feedback—in this case warming leading not to further warming but to its opposite, cooling.

This is because, as geochemist Wally Broecker of Columbia University's Lamont-Doherty Earth Observatory, has pointed out, when warm surface currents mix with fresh water such as that water already melted in the North Pole, the net salinity of the water, and therefore its density, is lowered. This prevents what would have been salty dense water from sinking and returning to the bottom of the ocean within the great natural conveyer-like circulation. Broecker discovered the interconnected system of ocean currents of which the Gulf Stream is a part, called the Great Ocean Conveyer Belt System. The ocean's surface currents are driven by wind but also move vertically. When the flowing ocean water evaporates in the sun in the tropical Atlantic, it makes for warmer, saltier water, which goes west and northeast to Europe. There the warm salty water encounters the cold winds that blow off Canada and Greenland. The cold salty water is very dense compared with other water so it sinks, and the natural cycle, which dissipates heat, continues.

Life when it is functioning well tends to do the same thing. But external events can disrupt the natural flow. If, for example, human life adds enough carbon dioxide to the atmosphere to melt polar ice, scientists expect the Gulf Stream to shut down. Even at a global level such hampered energy flow may indicate dysfunction, analogous to the fever and chills of a sickly person. And global cooling resulting from global warming is not so theoretical. Scientists know this because some twelve thousand years ago, in an event known as the Younger Dryas Event, the Gulf Stream shut down. Ice cores taken from Greenland show that the abrupt cooling event lasted about seventy years. The Younger Dryas Event is one of two relatively recent abrupt cooling events known from these ice core records. What seems to have happened in both cases is that a huge amount of glacial meltwater poured into the ocean from the St. Lawrence River in Canada after natural dams broke. The untoward result of the influx of polar meltwater was to shut off the Gulf Stream. James Lovelock predicts that Great Britain will become desirable real estate if nothing is done about global warming. Although the island nation will heat up, like other areas, the shutting down of the Gulf Stream, due to the intrusion of glacial meltwater into the North Atlantic, will compensate. Normally the Gulf Stream keeps England warmer than it otherwise would be. But with a dysfunctional Gulf Stream, the tropical waters carried past Florida will no longer arrive. The hotter Earth will not hit England so hard because the normal cycling that keeps the UK unusually warm for its latitude will be gone. But this meteorological blessing may be a curse for those already fretting over immigration to the heart of the empire that used to boast of being free from the setting sun.

Americans should perhaps worry about not just the waters flowing past Florida, but Florida itself, which is mostly below sea level. Waves may reach the torch held by the Statue of Liberty. Future geologists, if any, may unearth sunken condos filled with exercise equipment, plasma TV sets, and other items dating back to the maximal population of a certain bipedal species that was globally distributed in the early twenty-first century.

I'm interested in these funny feedback cycles. In a way, they defy logic. You're hungry, therefore you eat—whereupon you're no longer hungry. If we eliminate the intermediary and look only at the preliminary and terminal term, the bio-logic becomes: *I'm hungry . . . therefore I'm not hungry.* Novelist John Fowles in his slim metafictional novella *Mantissa* explains, in the words of his muse, "I am a woman. I am a tissue of contradiction." The point is that in cybernetic feedback systems stimuli can lead to a response opposite to the predicted response. Because the thermodynamic system is away from equilibrium, it may not react the way you think. The connected aspects of agency are interlinked, such that a tiny action may have great effects. The thermodynamic, the cybernetic system, acts alive: If you touch it, it jumps. This behavior, which we see in computers and in the reactivity of the global marine current system, also applies to ourselves.

Here's another weird idea. Our thirst ensures that we carry and integrate the compound in which our cell ancestors first evolved, before their descendants moved to dry land. Our thirst, in other words, is a form of life's lust for the conditions necessary for its continued existence. And because we are made (although we are really more "grown" than "made") of water, we require sufficient amounts to keep ourselves functioning as thermodynamic systems.

With two atoms of hydrogen and one of oxygen yet able to take forms ranging from steam to snow to ice, water was Thomas Huxley's favorite example of the whole being more than the sum of the parts. It seems ironic that *emergence*—now a scientific description of novel complex phenomena untraceable to their mere parts—

comes from Latin words meaning "to arise from the sea." Like Botticelli's painting of Venus, the goddess of love, emerging on a half shell in the ocean, water is a marvelous combination.

Because hydrogen (*hydros* is Greek for "water") is the lightest, most common element, and the one from which the nuclear furnace of the sun formed, it tends, in the absence of massive objects, to escape into space. When the sun and planets were forming, the inner solar system was richer in hydrogen. The Earth is only four and a half billion years old. The sun is not much older, and life is not that much younger. Preserving the elements of its origin, life is a kind of time machine. Our chemical cycling has preserved our watery, hydrogen-rich past. When you look into your friend's eye, you are closer, in a way, to the environment of the early Earth than you are to Rome. Life keeps going what has been going. It is a master of wet repetitions, such as sublime kisses, not to mention the slippery smacking and bodily fluids produced during the more profane act that sometimes follows. While it might be offensive to a social conservative, life is literally conservative. Indeed, so much does it repeat itself that author Arthur Koestler described evolution as an epic tale told by a stutterer.

Life's conservation of the conditions of its existence may extend to the character of water on our planet. Hydrogen has a stable isotope known as deuterium. Heavy water contains hydrogen in this heavier, isotopic form, which is exactly like other hydrogen except its nucleus comprises a proton and a neutron rather than just a proton. What is curious is that meteorites and space material in general are rich in deuterium relative to matter found on Earth. Like all of the isotopes of other elements, deuterium is identical to hydrogen except for its atomic weight. Its nucleus is heavier, and it has twice the mass of hydrogen. Heavy water, D_2O, is about 10.6 percent denser than ordinary water; ice made from it sinks in water. Heavy water can cause cell division and sterility, as well as bone marrow and gastrointestinal lining failure in animals. Bacteria, though hardy enough to grow in pure heavy water, grow more quickly in good old H_2O.

It is possible that life's cyclical processes have retained the early solar system environment in which water had a higher concentration of hydrogen to deuterium. In the thin Martian atmosphere of today, for example, water has a deuterium-to-hydrogen ratio that is five times higher than is the case for water found on Earth. Scientists have attributed this to the escape of hydrogen from Mars's atmosphere over time. Lighter than deuterium, hydrogen escapes more easily, leading to the relatively enriched levels of deuterium found in the Martian atmosphere and space debris generally. But the fact that life is so steeped in water in both its composition and its metabolism suggests that it may have something to do with the relative lack of deuterium and heavy water found on our home planet. Although speculative, here is more tantalizing evidence that Earth's ancient regime of life connected to natural cycles has literally altered—or rather kept unaltered—the face of our planet. It is a well-known biological fact (exploited in geochemical analyses) that when life takes in carbon, it chooses between taking in carbon 12 and carbon's heavier, stable isotope, carbon 13. Statistically life takes in more carbon 12. As a rule, life prefers the lighter elements, whether carbon, sulfur, or nitrogen. The same phenomenon, true of hydrogen, may have led to our happy, deuterium-depleted planet. Life, in other words, has a physical nostalgia. It hankers for its own. And this hankering, it seems—this wild thirst—has kept our planet wet.

Harvard University physicist Sheldon Glashow told a reporter: "The kids have to learn to love the world and to want to understand it, to be amazed at the stars and to be amazed at the wind and to see the wonder of the rainbow and see the changing colors of the leaves—to really want to know how these wonderful things happen, not just sit back and take it all as a magic show: but to try to want to figure out how the tricks are done. The difference is between just sitting back and enjoying or really wanting to know how it's done." Glashow shared with Steven Weinberg and Abdus Salam the 1979 Nobel Prize in Physics for their contributions to the theory of the unified weak and electromagnetic interaction between elementary particles. His comments on science and education were made during an interview with a local newspaper in Brookline, Massachusetts. I have done magic shows for his children during their birthday parties at his home in Brookline. Small world: He is my uncle by marriage, having wed my mother's next youngest sister, Joan. I certainly agree with the notion that we should not take things as given, but should try to fathom their hidden depths. And yet, as a magician, I've noticed that it's difficult both to appreciate a magic trick and to try to figure it out. When you wipe the dust from the mirror of your mind, even explanations can seem miraculous. Which reminds me of the magician Frank Garcia, who did a neat trick of strapping a coin beneath a rubber band on one side of a wooden paddle, which he then turned into two coins, doubling his money, and then, multiplying it further, a folded bill. Rhetorically asking the audience how he did it, he said, by way of explanation, "It's all done with mirrors"—flashing the paddle's newly reflective surface back and forth into the shining eyes of a bewildered audience. But of course, at least in magic, if not science and religion, something doesn't come from nothing. In fact, in a book devoted to the psychology of magic by one Dariel Fitzkee, the author reminds us that the general method for making something appear, or rather appear to appear, is not through magic, but through the removal of a secret cover.

The word *life* covers the thing, which is, it seems, like water. As evolutionist Richard Lewontin has pointed out, scientific devices that measure mist detect around each of us at all times a gaseous envelope that, despite our feelings of being delimited by our skin, must be considered part of us. Attached to water molecules travel scents and perfumes, the olfactory cues that bring back, beyond words, memories. Such strong, smell-cued memories may indicate the psychological side of our physical anchoring in an ancient aqueous world: Before there were words, before big-eyed mammals scurried in the night, before there were animals at all, there was life and water. And in the medium they didn't notice, because it was as good as part of themselves, organisms communicated chemically. Our sense of smell is a vestige, maybe, of a cell world of smell and sensation, primitive responsiveness and emotions unfettered by verbal clutter or squiggly marks on a printed page.

The human body
is composed of it, as are the bodies of our dearest pets, weeping willow trees, and ladybug beetles. Below zero degrees Celsius it can burn and can be hard and sharp, painful to the touch. It can hang in the muggy air in the dampened heat of midsummer. Its vapors can inhibit our own sweat. It can go unnoticed and invisible, or fog up our glasses with a mysterious conspicuousness. I see it in your eyes. Shape shifter and color changer, it is also the medium of connection of would-be separate land-based life-forms, which require its ancient properties to re-create the organic nexus of the ocean of life's dawn.

As a universal solvent, small changes in it may vastly alter its hue and appearance. As the evolutionary medium in which life evolved, it is a place to which we long to return. We pray for it, we yearn for it. Farmers and medicine men orchestrate dances and almanacs to wield power over its caprices. Models wet their whistles and continuously imbibe it to keep smooth and supple their sexy skin. It is the substance of tsunamis, the catcher of light in Turner's oils, the buoyant medium giving its liquid lullaby below as well as wrecker above of the sailor's skies.

Lubricator of Lolita and her narrator's lubricious tongue, taking its slippery trip down the palate to tap, at three, on the teeth, water is the wonderfully protean, clearly impure "metaform" that lends its slippery substance to smoking Nabokov and other life forms, chthonic and cosmic. Water, if it were not so common, important, and present, might have been called "dihydrogen oxide," its two atoms of hydrogen and one of oxygen, comparable, in a sort of reverse way, to silicon dioxide, SiO_2, that earthy stuff that also takes a variety of forms: jasper, tiger's eye, agate, rose quartz, heliotrope, onyx, flint, chert, glass, and sand.

The entire cosmos does not comprise mainly water, but one of water's ingredients: hydrogen. So does life. The hydrogen of the universe appeared from the beginning in the Big Bang some thirteen billion years ago. Hydrogen, the basic stuff of the entire visible sky, converted to helium and other heavier chemical elements in the center of those natural nuclear reactors known as stars. We recognize hydrogen on Earth mostly in its combined form as the elemental component, its atoms happily trapped in liquid union. Yet some hydrogen on Earth does persist in a purer form; it is a colorless, odorless, lightweight gas (H_2) that, as in a balloon, easily escapes from Earth's gravity. Life, when active, is always composed mostly of water. But life's ubiquitous H_2O, water, is not the elusive H_2 (hydrogen) gas.

On our arm of the galaxy, the sun and its companions ignited from a cloud of gas concentrated by gravity more than four billion years ago. At that time most of the hydrogen atoms—those that failed to remain in the sun—escaped to the outskirts of our solar system. Today hydrogen exists mainly as cold gas and hydrogen-rich ices (of methane, CH_4, and of ammonia, NH_3) of the outer planets (Jupiter, Saturn, Uranus, Neptune) and of their moons that retained gaseous atmospheres. But here in the inner solar system the bodies of the rocky planets—Mars, Earth, and Venus—have not been massive enough for retention of this atomic constituent of water, this lightest of elements. Hydrogen, the light stuff of stars, mainly has escaped to outer space from Mars and Venus, our planetary neighbors. Here on Earth, hydrogen lies hidden in one massive disguise; its abode is the three-thousand-meter layer of water on the surface of our planet, since the average depth of the world's oceans is three kilometers. Mars, too cold, and Venus, too hot, both lack any open bodies of liquid water and have retained, on their surfaces, less than a single millimeter of water as vapor or ice.

Although our three kilometers of hydrogen have had billions of years to escape, they have clung to Earth for a reason: the incessant thirst of highly active living matter. Life, which originated in water, remains composed of water. It needs what it is. Cyclically making and remaking itself, life may be as much the reason that water remained on Earth as water is the reason life remains on Earth.

As I sit here

and write on the warped, cracked, sun-bleached, and weather-beaten top of a picnic table, I think of this book as the cosmic equivalent of a message in a bottle thrown into the sea of space and time. Who am I writing for? Another? Our progeny? Myself in the future? What folly to address the future! There is no future (". . . no future, no future, no future . . . *for you*" to quote punk rocker Johnny Lydon, aka Johnny Rotten, who is said to have applied, unsuccessfully, for a marine biology program at MBL, the Marine Biological Laboratory in Woods Hole, Massachusetts). Not only because the future never technically arrives, not only because of what Paul West calls the "immensity of the now," but also because it's not looking good, climate-wise. Instead of wasting time and energy putting pen to paper in the lost cause of trying to warn a species that has only ever learned from experience, I visually drill down to the annual layers of growth outlined like white ripples upon brown water of the beautifully decrepit table—which I will in a moment be leaving due to being accosted by a bouncy-armed man whose speech suggests an unholy early afternoon mix of alcohol and depression medication. Like the displaced strata of this picnic table, caught on the way to decay, we write not only to taste life again, to relive the past, but also to send cryptic messages to ourselves or our descendants in the future. Before he fell from a mountain climb, physicist Heinz Pagels speculated that the laws of nature themselves might be a code left by our ancestors. The future is a mock foreplay, an unendurable fantasy infinitely prolonged. The future is the focus of our attention, the quenching liquid that recedes to an illusory oasis beyond the parched lips of Tantalus. The future appears now like a shadow of something huge and unseen, something that casts its dark light on the present, giving us glimpses of the shape of things to come, things that will have made sense, if at all, only in retrospect.

Why write? Just as life's squirming past leaves its traces in Earth's crust, so the here and now, exploding simultaneously and surreally with the reality of possibility and possibility of reality, is all that we have ever had. Why record it? It records itself. The authors of the Good Book were right: Nature begins as text as well as earth. It is silly to write because everything is already written, at least to someone who truly knows how to read.

In several of Philip K. Dick's science fictions, including the short story "We Can Remember It for You Wholesale," from which the film *Total Recall* was made, an amnesiac character leaves notes for himself that he dutifully recovers, not knowing their origin, in the future, when he studies them perplexedly. The facts of consciousness, our growing scientific knowledge and technological capabilities, suggest to me that there may be reality behind the scenes Dick stages in such stories. Is it possible that what we once knew will someday be fully recovered? That, perhaps having left ourselves the cosmic equivalent of Dickian notes along the way, we are slated to re-achieve—perhaps not in our own lifetime, but in the lifetime of the species, or of evolution itself—the original state of total awareness? Hindu thought develops the notion that Brahman, or universal consciousness, is reflected in Atman, or individual consciousness. But the notion is not devoid of scientific veracity. You and the universe are, whether your culture has instilled it in you or not, intimately related. The preponderance of hydrogen atoms in the water of our waterlogged bodies testifies to that fact of life.

From sparse tribal nomads, we have gone to town, or megalopoli: from nearly nothing to six-plus-billion humans in record time. Record time being our Holocene allotment, so far, of eleven thousand years. That's a lot, not of years, but of people. And with our massive numbers and lord-like powers of technology, energy use, and apparent dominion over other life-forms we feel that we have arrived. But where have we arrived? Only where we already were: on this blue, cloud-flecked planet, which, were it not for our bias as land animals, we would probably call "Water."

We have invented agriculture and, arguably, civilization. When asked what he thought of Western civilization, Mahatma Gandhi said, "I think it would be a good idea."

But ideas may have little to do with the formation of mass collections of living beings with powers beyond those of their individual members. Mass life-forms in the past have come together in their watery medium and merged, not only temporarily in sex but also permanently in symbiosis. Our two sets of chromosomes, one from each parent, recall our origins of being in a state of cellular doubleness. Biologists call this diploidy, and it probably first came about in the primeval muds. Why would a perfectly good reproducing cell with a nucleus and chromosomes bother to merge with a watery other? As we know from present-day relationships, it only complicates things. The answer, as often in science and magic, is probably disappointingly simple. Desire, the urge to merge, was an offshoot of a more primordial need, the requirement to have enough water and food to survive. The original lust that led the primitive cells with nuclei to double was probably more thirst than sex drive since, were it not for such cannibalistic behavior, our ancestors would have dried out and died—as no doubt did many of their brethren who did not get lucky. In other words, in times of drought and starvation, our cell ancestors—who had no immune systems to stop such things—probably hooked up because of hunger and thirst. In some cases the little devils digested each other's cytoplasms but not each other's chromosomes. Microbiologists have observed similar events—among hungry single cells that don't reproduce sexually—in modern laboratories. Although it is often stated that sex must exist for some reason, such as adding genetic variety, this suggests that sexually reproducing organisms have the option to reproduce nonsexually. But often that is just not the case. Yes, some species have closely related forms that reproduce without sex and fall prey more easily to disease; but such forms still undergo sex-like processes at the cell and behavioral levels. For example, although called asexual, the all-female whiptail lizards of the southwestern United States continue to mount each other: Their asexuality is not a true alternative, a clean break from sexual reproduction, but a curious form of it. Such populations of organisms may both be rare and lack staying power over evolutionary time.

Cycles of having two sets of chromosomes in the cells of bodies (forty-six in ours), and one set (twenty-three each) in the cells of sperm and eggs, are very old. They more likely testify to a series of genetic accidents, born of ancestral hunger and thirst, than a specific evolutionary advantage such as adding genetic variety. It was a case of do or die. Our ancestors did.

Inside the single cells that merged transparencies for the first time—engulfing each other and satisfying a more primitive urge for food and water—there were already living traces of other deeper, more permanent mergers. "I contain multitudes," wrote Walt Whitman, with perhaps more reality than he knew.

Assuming they breathed oxygen, a recent addition to the atmosphere at the time, our single-celled ancestors would have been dotted with mitochondria, the organelles that process oxygen. Genetic and other similarities of mitochondria to modern-day forms strongly suggest they are long-trapped oxygen-breathing bacteria. To this day mitochondria have their own genes, outside the nucleus, and reproduce on their own time table. They go wild, for example, in cancer. They can also reproduce as a result of exercise, which provides them with oxygen. It is generally accepted today that the mitochondria are bacteria that invaded or were eaten by larger, amoeba-like cells. Like sex, only more lasting, here was a living union, long before civilization, with serious consequences for life.

Today bacteria breathe oxygen—and methane, and hydrogen, and hydrogen sulfide, and arsenic, and much else besides. But our lineage and the pedigree of almost all visible life-forms from mushrooms and seaweeds to sunflowers and apple trees is from the merged sex cells that themselves incorporated respiring bacteria. The respiring bacteria incorporated into the amoeba-like cells in more familiar beings were themselves mutants. The breathers had spread in the aftermath of the biggest pollution crisis in Earth's history—the buildup of oxygen in the atmosphere that is chronicled in records of rust, uranium oxide, and other compounds determined by radioactive dating of uranium and lead to be about 1.8 billion years old. Geologists will date anything, as you may have heard.

Where did the oxygen come from? Well, as you may have guessed, the water users put it there. And who were the water users? They were the single cells, bead-like strings, and larger colonies of teeming blue-green algae that, though most science writers and scientists call them that, were not algae at all but rather the aquamarine bacteria properly known as cyanobacteria. The cyanobacteria were themselves mutants. And they had hit upon the literally Earth-changing trick of using water as a source of hydrogen. All life needs hydrogen to build its bodies. We get it from food, but we can't make our own food. Cyanobacteria can. Using light, they combine the hydrogen of water with the carbon of carbon dioxide in the air. Other organisms eat them. We can, too, although we generally feed higher up on the food chain. Today cyanobacteria, and their formidable oxygen-producing descendants—plants—forge the links of that chain.

The ancestors of green beings responsible for adding free oxygen gas (O_2) to Earth's atmosphere probably took their hydrogen straight, as some bacteria do today. Metabolically similar purple bacteria now breathe volcanically burped hydrogen sulfide. Their numbers may have declined when hydrogen, escaping to space, became scarcer and volcanoes became less active.

The green mutants grew with wild abandon, emitting oxygen into the atmosphere for millions of years. The free oxygen gas took a long time, reacting with the vast surface of the ocean, and then the rocks of the land, before it accumulated to its present concentration of roughly one-fifth the atmosphere.

Moreover, the green mutants were themselves symbiotic, like the breathers that derived energy from their waste. Today plant cells have little green inclusions called chloroplasts that do the actual photosynthesis. Again, genetic and other evidence ties them to free-living bacteria. The cells of an adult animal generally have two sets of chromosomes, as well as mitochondria that come from the ancient days when oxygen-tolerant and then oxygen-using bacteria spread over the face of an increasingly energized planet. The cells of plants generally have not only mitochondria and two sets of chromosomes but chloroplasts as well. The sex merger, the breather, and the green ghost.

From kelp-like ocean flotsam to light-seeking stem beings, the water users and their hangers-on evolved into jungles, dense ecosystems of plants and the animals they attracted with their bright colors and exotic fragrances. They clothed us and fed our kind. In return, wittingly or not, we disseminated their seeds. Life is not just human but layers of trapped space-time, cells within cells. Beyond, above, below, and outside of civilization, energy-using, energy-storing life has been subtly organizing collectives since long before we evolved. In fact, as the foregoing brief history of water suggests, intramural activities were responsible for the appearance of sexually reproducing cells and animals—us—in the first place. Geologists, paleontologists, and cell biologists together reveal the past. They strip away the secret cover of the present, revealing a fuller picture of time. Such a picture, though incomplete, makes more sense than the one provided by human culture alone.

When we contemplate
the future, we cannot simply extrapolate technological trends indefinitely. If we did that, then historical increases in travel speed from horse-and-buggy times to supersonic jet would continue exponentially beyond the speed of light. Because the measured speed of light is a finite upper limit, the absolute boundary that frames relativity theory, this seems unlikely. Moreover, our exponential increase in population must halt, as it has already begun to do. Populations of bacteria in petri dishes are sometimes most abundant in the generations preceding their collapse. Hopefully this is not the case with us. Our global distribution, and the possibility that we will colonize other worlds, increases the chances that at least some of us will survive a catastrophe, whether from the outside such as the planet being hit, as it is every hundred or so million years, by a massive meteorite, asteroid, or cometary nucleus, or from the inside, by life's own toxic pollution, infectious mutations, or war making.

In *Galapagos* by Kurt Vonnegut future humans are whittled down to only a handful by a combination of world war and a global pandemic virus. The lucky survivors are an odd group of pampered tourists visiting the Galapagos Islands. They survive idyllically on the beach eating fruit and whatnot but are in danger of dying out until an old man, taking humanity's future into his own hands, literally, secretly introduces his semen into a couple of the younger sleeping females. In due course mutations occur, and the inbred progeny of the great human technological experiment revert to a more tried-and-true evolutionary form: The blubbery offspring take to the water. Their bodies become hydrodynamic, and their heads are so pointy that they can jump in the water without making a splash. After a few generations, Vonnegut explains, the only thing connecting them to their human ancestors is that when they are lying out in the sun, and one of them farts, the rest of them laugh.

A fitting end, no doubt, and one that moreover accords with the history of other land mammals, such as the ancestors to walruses, dolphins, and whales, which have returned to the water. Such evolution by different species in a common direction is called convergent evolution. English paleontologist Simon Conway Morris speculates that the presence of convergent evolution suggests that intelligent Earth life is no accident, and that human-like beings can be expected to arise along with Earth-like life on other planets. The future, known only in retrospect, in "the immensity of the now" remains elusive by definition. I was walking into the bowels of what is reportedly the tallest library in the world, the W. E. B. Du Bois library of the University of Massachusetts–Amherst, when I noticed, as part of a special exhibit, illustrations accompanying the "golden age" of classical science fiction. Plastered with logos of the then-icon for the atomic age, aloof electrons ellipsing their dangerous nucleus, the artwork, strewn, like the written work itself, between the comic and the surreal, took hold of my imagination with its crew-cut captains, rocket ships, and extraterrestrials. One picture showed Isaac Asimov (whom my mother always said was arrogant and proud of being alphabetically first with his A, and whom I saw for the last time on the top floor of the World Trade Center during my father's wedding), the elegant L. Sprague de Camp and Robert Heinlein (whose circular house I once visited as a boy) posing together in 1944 at the naval Air Material Center of the Philadelphia Navy Yard. Covers of books by A. E. van Vogt were also on display. Van Vogt's powerful imagination laid down tracks for the future of science fiction. His two-hearted, large-brained Johnny Cross, with telepathic tendrils, the mutant hero of *Slan,* was a precursor to Marvel's X-Men. Van Vogt's imagination helped inspire comic artists Jack Kirby and Stan Lee in their Fantastic Four and New Gods creations. Ridley Scott, director of *Blade Runner,* was a fan, and admits the film *Alien* to have been inspired by a van Vogt story. Philip K. Dick also recognized van Vogt as a formative influence.

In one van Vogt story, two spacecraft rendezvous. They link hulls, breed their human inhabitants, and, having accomplished their mission, continue on their interstellar way. Literally, this is a far-futuristic post-human scenario. The spacecraft has become the skin or exoskeleton of the human. The mode of reproduction has radically altered, and with it the phenomenology of human existence. How strange. And yet this is, I would argue, an allegory for what has already happened to people. When we look at these lonely, far-future descendants who have forgotten their past, but occasionally link up for some quality time in the depths of space, we have to think of ourselves. The events that induce us to seek each other out and regenerate new beings that grow from amoeba-like cells are part of a bizarre history. Insofar as "we" remain cells, the "spaceships" of our bodies are at once glorious craft carrying us across dry land and cyborg extensions of our soft and slippery inner core. When Neil Armstrong stepped on the moon, he had mitochondria (processing oxygen from his air tank) in his foot. In the same way, if humanity somehow survives to colonize space, it is very unlikely we'll do it alone. More likely we will be like the mitochondria, useful parts integrated into larger technological and biological systems. One giant leap for humankind, it will be the next logical step for growing, energy-transforming, naturally expanding life.

The incorporation

of multiple consciousnesses into a single individual is a recurrent theme in science fiction. As a teenager I stayed for a few days in the company of science-fiction author Theodore Sturgeon, the real-life model for Kurt Vonnegut's fictional character Kilgore Trout. His novel *More Than Human* was about a group of human misfits, carnival characters really, who merge on another plane of being to have powers beyond any alone. In *Childhood's End,* Arthur C. Clarke portrays a bevy of children progressively losing their ordinary behavior and coming together to make contact with or as a new kind of being. So, too, in *The Three Stigmata of Palmer Eldritch,* the drug Can-D is consumed by otherwise deadly bored human migrants to Mars. In this Philip Dick novel the Martians live in hovels and, putting their Barbie- and Ken-like Perky Pat dolls on special layouts, merge consciousnesses, hallucinating that they are together inside and seeing outside the eyes of the toys, which has the effect of distracting them from their extraterrestrial squalor to the point that they believe they are on a sunny beach with tan young men, cars, and girls.

The merging of consciousnesses into an "individual" seems bizarre. And yet, is it? If we grant awareness at some level to the prokaryotes that merged to become the eukaryotic cells that, multiplied in the process of growth, evolved into animals, then we may wonder whether our own individuality, so "real," is not stranger still than the fictional fruits of a novelistic mind given to paranoia and running to overdrive on methamphetamines and sleepless delusions. The drug Can-D miniaturizes one's sense of spatiality to the point that the pill-popping party imagines herself part of a unitary consciousness within a Martian toy. Even without the foregoing speculation that our consciousness is partially the result of merged symbionts, we should recognize something familiar here. In a way the most bizarre scenario Dick dramatizes is as commonplace as teatime among the English gentry. In fact, you may be experiencing a version of it as you read. Dick reframes the ordinary as the extraordinary, blurring its edges until its source may not be recognized. A number of people psychically inhabiting the bodies of a smaller number of characters, in a different place and time—haven't we seen this before?

For the drug Can-D, with its strange pooling of psyches and embodiment in human form, resembles precisely what the reader is doing while engaged in the hallucinatory act of reading. In perusing the ink spots upon the bleached canvas of sliced trees, we, too, hallucinate, we mentally enter the mind of a character, see his hat or her red dress, look at the landscape through the eyes of another: We mutually project our perception into imaginary beings laid out in words.

Moreover, as Plato writes in *The Phaedrus,* writing is a drug. Embracing the concept even as he distanced himself from it by putting it into the mouth of Socrates, who never wrote, Plato calls writing a *pharmakon,* Greek for "drug." Like the English word, the term *pharmakon* has a double meaning: It can be either a poison or a cure. For Plato's Socrates writing is toxic, because it replaces the oral tradition and the use of one's own mind, especially one's own memory. But of course it is also helpful—augmenting our powers of memory once we consider them not to be confined to our head, but to extend to the library and philosophical literature. In the Seventh Epistle, a Dutch philosopher told me, Plato suggests that unified consciousness, the meeting and melding of minds, is the numinous experience par excellence, the closest we can get to God. Life is literally con-fused. In conversation, in performances, in companies and personal relationships new ideas come that never would arise for an individual alone. Such emergence is intriguing.

Life is chimerical. The individual is both an aggregate of organisms and a part of an aggregation. The organisms making up individual plant and animal cells and bodies were not originally, nor are they still in many cases, of the same type. The ancestral bacteria may have even been hostile—invading and eating each other before establishing various degrees of tolerance, harmony, and disharmony. But over time, odd couples—and ménages à trois, foursomes, and still higher-numbered partnerships—triumphed. In Isaac Asimov's *The Gods Themselves,* three sexes dwell upon the surface of a distant planet. The mutually attracted organisms interact fleetingly if flirtatiously, floating through one another like extraterrestrial ghosts until they reach physical maturity. When that time comes they mate permanently and metamorphose into a "hard one."

These humans from Earth

raze forests, spread farmland, and turn liquid in the ground into gases in the atmosphere. Like the oxidizing microbes before us, we are changing the world. Like them, we are, simply because of location, among the organisms most threatened by our activities. Our addition of carbon dioxide to the atmosphere, even if it does melt the polar ice caps, drown the beautiful polar bears, and erase what is left of human civilization, will be paltry planetary damage compared with previous global crises generated by asteroid and meteorite impacts, by extinction-linked bouts of subsea and surface vulcanism, and by rampant growth of mutant bacteria. It is amusing that, as soon as we lose the biblical arrogance that persuades us that the entire Earth was made for us, we overcompensate with the opposite delusion that we are so dangerous, all life is imperiled by our actions. Both of these egotistic views, although opposed, are highly unlikely.

Paleontologists have identified five major mass extinctions, including the relatively recent one that saw the rest of the non-flying dinosaurs in the wake of a bolide—an exploding fireball that blasted into the surface of the Earth, settling in the ocean beneath the Yucatán and forming the Chicxulub Crater beneath that town on Mexico's Yucatán Peninsula. The flaming object, which left traces of a rare Earth metal, iridium, in a thin layer of clay around the world, may have been a rock-and-metal meteorite. Meanwhile—3.88 billion years earlier—a glancing collision with a planet-size object called Theia may have formed the Moon. Earth has been submitted to extreme violence, both external and internal, from the time of its origin more than four and a half billion years ago.

By some taxonomic grouping measures the Permo-Triassic and Triassic-Jurassic extinctions were more severe than the more famous Cretaceous bolide event. These other events, leading to multiple extinctions of plant and animal species, may also have been caused by extraterrestrial impacts, although the evidence is not so clear. It is possible that volcanic events, perhaps triggered by collisions, were involved. But not usually counted on the list of mass extinctions, numbered by zoologists at five (with humans causing the "sixth"), is the trail of death left by ancient bacteria. The cyanobacteria, which mutated some two billion years ago to use water as a source of hydrogen to build their bodies, were responsible for a catastrophic influx of oxygen molecules into the atmosphere. Oxygen gas went from nearly nothing to almost one-fifth of the atmosphere: The result was that the remaining life, if it didn't retreat to anoxic muds, or develop ways of detoxifying dangerously reactive oxygen, perished.

On the other hand, the survival of green water-using life-forms—and some bacterial strains thrive today inside nuclear reactors—provides philosophical consolation. The ordinary person wants to save the Earth. But she doesn't mean save the Earth without humans. When she says *Earth,* she means the system that supports humans. This system was in place long before humans, however, and will likely go on long after us. It has survived wholesale metabolic regime change—the switch to an oxidizing atmosphere—not to mention extraterrestrial impacts releasing more energy than would be the case for an all-out nuclear war.

During the sweltering Eocene there were crocodiles in the Arctic circle. The Eocene heat wave lasted less than one hundred thousand years—about ten times the length of the Holocene and roughly of history itself, considered (as do historians) from the origin of writing.

We know that life can screw up the global environment. Cyanobacteria screwed up way worse than we did when they added free oxygen to the atmosphere starting some two billion years ago. Then again, a lot of death ensued before organisms evolved that considered their waste fresh air. Although we neither respect nor notice them, the water-using Earth-changing cyanobacteria survived. So far we have, too. But unless we get our act together, or the observed current global warming turns out to be caused by some natural phenomenon, such as our solar system swinging nearer one of the spiral arms of our Milky Way galaxy, we may not be so lucky.

A steamy planet

may not support human beings in anything like our present numbers. Typically, when an invasive organism reproduces in a living environment, it disrupts the physiology of the host, sickening it. We all have had the experience of "chills"—fever alternating with feelings of sickly cold as our bodies work overtime to stabilize our temperature. The same thing can happen on a planetary scale. Sophisticated but unconscious, like our pumping heart, the biospheric environment is autonomic—performing highly complex cycling and interspecies physiological functions without our having to pay the least attention. But then we don't pay the least attention to our heart. Unless, of course, it malfunctions. Fortunately, visionary scientists are newly recognizing the autonomic structure of the planet—which may experience immune-system-like responses, including global chills, as it unconsciously tries to stabilize its fever. The age of the universe is currently astronomically reckoned at thirteen billion years. A thousandth of that would be thirteen million, about three times the length of time since the appearance of even the sorriest, most hunched and hairy excuses for human beings on the African savanna. But while humans have existed in some form for at most a few tenths of a percent of the time since the Big Bang, the Gaian biosphere is one-third the age of the universe. "She" perhaps knows—at least at a bodily level—more than we. As when we drink too much coffee and get tachycardia, the Earth system has been *disturbed* to the point at which once smoothly functioning natural systems have been upset.

Although for some reason it wasn't picked up by the major newspapers, a recent press release disseminated by the Associated Press reports that a group of climate scientists from the Royal Society (of London), in conjunction with colleagues at the National Academy of Sciences (in the United States), made a formal statement to the United Nations—as well as sending individual notices to every recognized government on the planet. The missive reads, in part: "After decades of collecting and analyzing data, the homeorrhetic feedback mechanisms of our planet, as well as their partial biogenic nature, are hereby recognized. The signatories . . . have concluded that Earth's biosphere, is alive. . . . A genuine planetary macroorganism . . ." In their "Emergency Recommendations Protocol," the scientists urged:

> 1) The joining of nation states and international governing bodies in formal recognition of this *sui generis* macroorganism, 2) The immediate organization of an international conference on planetary health and multispecies function to set down strong recommendations for all future global economic and environmental policy, and 3) The convocation of financial, scientific, institutional, and technological resources so that the macroorganism may be carefully observed and cataloged using the latest satellite information and communications technology. . . . it is possible that this additional computing power, combined with the social-biological planetary integration of such a project, will permit contact with extraterrestrial life forms that have hitherto avoided us due

> to their shunning of unstable life forms . . . these edicts provide us with serious counsel during this perilous time in our global evolution.

The document even says that intelligent extraterrestrial contact with us may have been thwarted by our species' lack of explicit recognition of the grander organism of which it is a part.

An astounding document, to be sure. If you missed it, don't be too surprised. Such is the power of propaganda, of spin and willful neglect in our media-controlled age. Our political leaders are simply not ready yet to consider life from the planetary level of our most forward-thinking scientists.

True as that last statement may be, there was no press conference. I made it up. I got the idea from composer Nathan Currier in a letter he wrote to Earth system scientist Tyler Volk. Volk had written a review of Gaia founder James Lovelock's *Revenge of Gaia: Earth's Climate Crisis and the Fate of Humanity.* Currier, one of whose works is the somewhat screechy *Gaia Oratorio,* takes an interest in the grand debate of the biosphere's status as an organism. Of such an announcement, Currier whimsically wonders: "Might it not be greeted somewhat like the announcement of finding extraterrestrial life—indeed, *isn't* it in a way like extraterrestrial life, in that it is only clearly visible from the vantage point of space? Think of the potential of this to stir people, to change the debate, to create the political will to remake our economy from the ground up (based on the *ordre naturel,* as the Physiocrats suggested way back in the time when dear Mozart was but a child prodigy)."

For the moment the notion of a living Earth seems like hokum to many scientists, because referring to Gaia as a living thing with awareness or intent violates the scientific norm of considering only human beings to be endowed with genuine subjectivity. The rest of the world is a mechanism to be manipulated and observed, and if possible controlled by the scientific, technological mind. A living Earth is also an affront to those inculcated in traditional Western religion, which reserves consciousness, awareness, and full purposefulness only for human beings.

Another review of Lovelock's *The Revenge of Gaia*—"Goodness, Gracious, Great Balls of Gaia!"—appeared in the August 2006 online version of the magazine *American Scientist.* In it Brian Hayes writes that "Lovelock's . . . 'geophysiological' system is endowed with a will and a personality, not to mention a gender. She is . . . a 'tough bitch.'"

Although Lovelock provides a disclaimer that he is only using "the metaphor of 'the living Earth' for Gaia [so] do not assume that I am thinking of the Earth as alive in a sentient way, or even alive like an animal or a bacterium. . . . It has never been more than metaphor—an *aide pensée,* no more serious than the thoughts of a sailor who refers to his ship as 'she'"—and while Lovelock knows that "to personalize the Earth System as Gaia, as I have often done and continue to do in this book, irritates the scientifically correct"—he is steadfastly "unrepentant because metaphors are more than ever needed for a widespread comprehension of the true nature of the Earth and an understanding of the lethal dangers that lie ahead."

Hayes, admitting he is "scientifically correct," then says he is "sympathetic to the idea that the global ecosystem must be reasonably stable and self-correcting. But self-regulation requires no purposeful, animating spirit; Lovelock's vision of Gaia twiddling the controls to keep us comfortable—or else deliberately turning up the heat to snuff us out—leaves me utterly baffled. Even as metaphor it's nonsense."

Well, nonsense next to what? Reasonable people, as we have seen, can logically surmise that the biosphere may have awareness; we have no way of definitively knowing, any more than we can glean whether a sufficiently complex computer program is "awake." In the famous Turing Test designed by the brilliant mathematician and World War II code breaker Alan Turing, he argues that if a computer truly persuades a respondent that it is conscious, we have no choice but to so regard it. If we apply a version of the Turing criterion to the Earth system we must, I believe, assume it is alive. The biosphere displays evidence of physiological behavior. According to Turing, if a computer answers questions to the satisfaction of an observer in a way that seems to demonstrate life, we must conclude the computer program is alive. We have no direct access to the subjectivity of others, but if we get the requisite signals we can empathize with another and ascribe it, him, or her true subjectivity. Basically, Turing is saying that if it walks like a duck and squawks like a duck, it is a duck. If a zombie convinces us he has feelings, well, then we must assume he has feelings. There's no difference between an original and a perfect copy. In the same way, in the case of the biosphere, if it demonstrates physiology, we are justified in arguing that it is alive. Indeed, before Lovelock invented his Daisy World model, the geological evidence for atmospheric chemistry and active global temperature regulation was dismissed because it seemed to imply an impossible-to-explain global version of mind. But these phenomena do exist. We are thus justified not only in considering Earth alive, but also in suggesting, as does Lewis Thomas, that it may have something akin to consciousness. It may even have a consciousness that dwarfs our own, despite how repellent we naturally feel that any being in the universe other than one made in our likeness could be smarter than we are.

More importantly, the notion that purposefulness lies wholly in the human realm is, although scientifically orthodox, an unnecessary excess of unscientific anthropomorphism. Nothing could be less "Copernican," nor even "evolutionary" in the sense of realizing our kinship to the rest of living beings—as well as to the energetic and chemical processes from which life, scientifically considered, comes—than to assume a priori that the qualities found in human beings are prohibited to the rest of dumb, mechanical nature. (I am assuming with the true scientist that life, whatever mystery in it still abides, should be studied as a natural phenomenon.) This is a philosophical point that Lovelock does not finesse—but then scientists are not only not trained in philosophy but actively tend to disparage it despite the fact that all science's assumptions, however reasonable and realistic, lie outside of science itself.

Assumptions always exist. The assumption of modern science in this case, associated with the thought-gestalt that everything is physical (sometimes known as metaphysical materialism or naive realism), is that human subjectivity, intent, thought, sensation, and goal-seeking behavior are rare or unique in the natural world. This is a dubious proposition in light of humanity's putative emergence from the natural world and life's presumed emergence from the nonliving world. Who are humans to deny *on principle* to others the experiences we have? As a child might put it, who died and made us king? In fact, we may be emperors without clothes insofar as we insist that our capacities to experience, to choose, to reflect, and to act with purpose are unique in the natural world. These all may be true but, then again, they may be mere assumptions. I think they are. Careful examination of even nonliving systems shows them to exhibit purposeful behavior, irrespective of the unverifiable possibility that nonhuman beings (or even processes) with which we cannot communicate have genuine experience.

Not surprisingly, given what science has learned in the last several centuries about humans not living in a special place (heliocentrism), not being separated from other organisms in either space (ecology) or time (evolution), and being composed of not just common but astronomically the *most common* elements (hydrogen, oxygen, carbon, et cetera), we good Copernicans should not be surprised that our orientation toward sources of energy to preserve our chemically complex being is also not unique. So when Hayes, a good hardheaded scientist, cannot help himself from calling Lovelock a "crank" in print because of his predilection, exercised with due caveats, to attribute intention to the would-be dead and mechanical biosphere, perhaps he should think again. One is reminded of the story of an interaction between logician Bertrand Russell and process philosopher Alfred North Whitehead.

"The problem with you, Alfred," said Russell, "is that you are wooly-headed."

"The problem with you, Bertrand," said Whitehead, "is that you are close-minded."

In my estimation the "problem" with Lovelock—apart from his lack of time or inclination to develop a more careful philosophical or rhetorical package to present his intuitions—is that he is not "close-minded." We should perhaps listen less dismissively to this institutionally unbeholden scientist who not only postulated global life's physiological behavior on the basis of thermodynamic data, but also invented the previously noted electron capture device that led to discovery of pesticide residues in Arctic animals and the role of chlorofluorocarbons in creating holes in the ozone layer. To speak of Earth's potential behavior in metaphors taken from human activity not only gets the point across but is indeed preferable to the bankrupt notion that Earth is a mere rock that happens to have some life on it. In sum, it may be nothing short of foolhardy to insist at this point—unless we wish to continue swimming in the religious deep end—that the only sort of thing that has true purpose, intent, and consciousness just *happens* to be human.

As cultural historian Richard Tarnas, discussing these matters in another context, has said,

> The systematic recognition that the exclusive source of meaning and purpose in the world is the human mind, and that it is a fundamental fallacy to project what is human onto the nonhuman, is one of the most basic presuppositions—perhaps *the* basic presupposition—of modern scientific method. . . . The quest to liberate the human being from the bonds of nature through human intelligence and will, to ascend and transcend, to gain control over the larger matrix from which the human being was attempting to emerge . . . climaxes in modernity, especially in modern science, where the dominant goal of knowledge is ever-increasing prediction and control over an external natural world seen as radically "other": mechanistic, impersonal, unconscious, the object of our powerful knowledge . . . speaking very broadly, we may say here that as the human self, guided by its evolving cultural, religious, philosophical, and scientific symbolizations, has gained increasing substantiality and distinction with respect to the world, that self has increasingly appropriated all the intelligence and soul, meaning and purpose it previously perceived in the world, so that it eventually locates these realities exclusively within itself. Conversely, as the human being has appropriated all the intelligence and soul, meaning and purpose it previously perceived in the world, it has gained more and more substantiality and distinction with

> respect to the world, accompanied by ever-greater autonomy as those meanings and purposes are seen as ever more malleable to human will and intelligence. The two processes—constellating the self and appropriating the *anima mundi* have been mutually supportive and reinforcing. But their joint consequence has been to gradually empty the external world of all intrinsic meaning and purpose. By the late modern period, the cosmos has metamorphosed into a mindless, soulless vacuum, within which the human being is incongruently self-aware. The *anima mundi* has dissolved and disappeared, and all psychological and spiritual qualities are now located exclusively in the human mind and psyche. . . . Must we not regard the interpenetration of human and cosmic nature as fundamental, radical, "all the way down"? It seems to me highly improbable that everything we identify within ourselves as specifically human—the human imagination, human spirituality, the full range of human emotions, moral aspiration, aesthetic intelligence, the discernment and creation of narrative significance and meaningful coherence, the quest for beauty, truth, and the good—suddenly appeared ex nihilo in the human being as an accidental and more or less absurd ontological singularity in the cosmos.

So let us not be too quick to judge in favor of our uniqueness or superiority. What we have may belong also to the biosphere. "She" may be a mite more sophisticated than us—and so may the universe itself. These really quite sane possibilities run against the current cultural grain, not only because of the methodological prohibition against subjectivity in nature standardized in modern science by people such as Descartes and Bacon, but also because of the special relationship that humans, and humans alone, are supposed to have with their

The feeling that you are at the center of the universe is a powerful illusion, and true insofar as it goes. But everyone else also has this feeling of being at the center of the universe. And everyone else is not necessarily human.

monotheistic God. Whatever else this relationship may be, it is not scientific. Scientists should be careful about pledging their unwitting allegiance to a deity that keeps us special, apart, and alone in defiance of the great Copernican trajectory of recognizing ourselves to be, on the basis of evidence and experiment, at one with the natural world.

I asked plant geneticist Wes Jackson how we should think of the biosphere, if not as an organism. "How should we think of God?" he asked. I didn't have an answer. I brought up the idea that, to me, if the Earth is alive, not just a chemical and a physical but also a physiological system, perhaps it is even aware at some level. Probably not, but it's possible: If we are "nothing but" a long-evolved colony of cells, and seem so superior to our cell constituents, what might a system whose parts consist of us be like?

Jackson suggested that the problem is one of pride. There is something to be said, he remarked, for the old idea of a great hierarchy of beings, with people above—and more responsible than—other animals, yet below—and destined never to be as powerful as—the angels. Such systems can keep us humble.

But so can thinking of the Earth as a living being capable of cybernetic, physiological, immune-like action that can put humanity—currently overrunning the planet like a global infection—in its proper place. From a planetary perspective, humanity may be quite valuable because of our technological prowess, which gives Earth life as a whole the chance to survive in space. But as global warming suggests, the sheer quantity of our kind and operations is beginning to damage the smooth thermodynamic functioning, dependent upon exporting heat and entropy into space, of the planetary whole.

Earth, from space, is not the static map we see in our flat books and atlases. Astronaut Eugene Cernan observed from his Earth-orbiting perspective that "You literally see North and South America go around the corner as the Earth turns on an axis you can't see and then miraculously Australia, then Asia, then all of America comes up to replace them. You see a multicolored, three-dimensional picture of Earth. You begin to see how little we understand of time. You ask yourself the question 'Where really am I in space and time?'"

Throughout our daily lives we move through the biosphere with all the awareness the proverbial goldfish has for the fishbowl water in which it swims. You can take a man off the Earth, but you can't take the Earth out of a man.

As my late father, Carl Sagan, put it, Earth from space is a "pale blue dot." This is what our watery dwelling looked like from *Voyager 1* when it took time out on its voyage beyond Jupiter and the solar system to photograph the planets it had passed, and the location of its origin:

> Look again at that dot. That's here. That's home. That's us. On it everyone you love, everyone you know, everyone you ever heard of, every human being who ever was, lived out their lives. The aggregate of our joy and suffering, thousands of confident religions, ideologies, and economic doctrines, every hunter and forager, every hero and coward, every creator and destroyer of civilization, every king and peasant, every young couple in love, every mother and father, hopeful child, inventor and explorer, every teacher of morals, every corrupt politician, every "superstar," every "supreme leader," every saint and sinner in the history of our species lived there—on a mote of dust suspended in a sunbeam.

So much depends on perspective.

Your body, wet on the inside, protected by waterproof skin on the outside, traps the conditions of the early Earth when the first cells formed. On land each of us walks in an invisible envelope of mist produced by our bodies and breaths. Whether or not other life-forms can exist without water, it seems clear that if our kind makes it into space, we will take wild watery worlds with us.

The desire to get off the Earth and live beyond our atmosphere in space has been called a modern religion. By making new biospheres—technological enclosures that recycle, with the help of life, their gases, liquids, and solids—we see how such space living might be done. But long before we see a New Earth erected on another planet by spacegoers, we may produce the lost island of Atlantis, not a barnacled ghost town but a thriving, electrically lit metropolis with ocean-based economies built on biospheric technologies.

Renowned physicist Stephen Hawking argues that our survival as a race increasingly depends on our ability to find new homes elsewhere in the universe. Just as, the longer you play at the gaming tables, the greater your chances of losing everything, there is an increasing risk that our planet will encounter a major disaster leading to human extinction. One possibility Hawking mentions is an AIDS-like mutant that is airborne, spreading by sneeze rather than sex. Then there is widespread nuclear war, not to mention Earth collisions with space rocks such as the one thought to have created the Moon or the more recent bolide crash that led to the extinction of the last of the flightless dinosaurs.

As Stephen Jay Gould has pointed out, species able to settle distinct locales, such as separate islands by swimming, will under certain conditions have an advantage over far more populous species that are not so migratorily able. If all members of a seemingly successful species, for example, inhabit a single island, and a tidal wave comes drowning them all, they will go extinct while a superficially more lackluster species, existing in small numbers but on several islands, will be more likely to survive. The same is true for humans and life generally on Earth: The more celestial bodies we inhabit, the better our chances for our long-term survival—because, once in space with life and technology, we may morph beyond the recognizably human into many new forms.

Hawking, better known for cosmological models of the entropy of black holes than his advocacy of space exploration, claims that humanity could have a permanent base on the moon in the next twenty years and a colony on Mars in the next forty: "We won't find anywhere as nice as Earth unless we go to another star system." In their story *The Greening of Mars,* Lovelock and Michael Allaby imagine a future in which missiles designed to carry nuclear weapons are used to carry fluorocarbons to Mars instead; although subsequent calculations proved the plan unfeasible, the idea of raising Mars's temperatures to a range that could support life, this modern version of turning swords to ploughshares, was a good one. It is also possible to imagine bombarding Mars to create craters so deep, atmospheric pressure at the bottom (eight to thirteen kilometers) would be similar to Earth's, as part of a preliminary step in the process of terraformation, a suggestion made by James Oberg in his book *New Earths*. Hawking believes that if we can avoid killing ourselves in the next hundred years, we should be able to set up space settlements free of the need for support from Earth. What this means, from a practical standpoint, is that not just technological humans but also enough other species to re-create ecological cycles must become part of the space settlements. A bleaker alternative—which may become technologically feasible—is for machines to synthesize food directly from carbon, hydrogen, oxygen, and other atoms in space. Even if this does prove possible, however, we might feel impoverished in a technological environment eviscerated of other life-forms.

"It is important for the human race to spread out into space for the survival of the species," Hawking has said. "Life on Earth is at the ever-increasing risk of being wiped out by a disaster, such as sudden global warming, nuclear war, a genetically engineered virus or other dangers we have not yet thought of."

An experience in a dream gave me an insight about space. On more than one occasion I have felt myself to be little in dreams, as if I were a little child. In this dream—and it is very tactile, very vivid, almost as if it were a memory rather than a dream—I lean back in my chair. Right at the part where I might expect a parent to tell me to sit up straight, something strange happens. I lean back farther, but I do not fall. The feeling of weightlessness that overcomes me saves me. Sometimes it is a prelude to floating in a sort of dream space walk. If normal gravity were operating, I would no doubt slam to the floor.

What is weird is the strange realism in the dream of the feeling of floating. I was left with this feeling, very close and un-dream-like, when I awoke. At first I couldn't imagine why it was so un-dream-like. I even remember thinking how real it seemed compared with most flying dreams—except perhaps for those rare flying dreams made so real because of the perfect simulation of wind resistance due to a breeze from an open window sweeping over one's sleep-stilled shoulder.

The source of the mysterious realism eluded me until it occurred to me that my dreamed feeling of not falling, but floating, may in fact have been a real one. For all of us have floated as embryos in the amniotic fluid.

That watery sac in the mother's womb may have left its memory in the infant's developing sensorium and memory. And, having left its mark, it may have enhanced my experience, in a similar state of stupor, of the dream's realism. It is ironic that the dim memory of the floating fetus—an image of the past—seems to anticipate the experience of the futuristic space voyager, freed, at least momentarily, from the clutches of gravity. In James Cameron's IMAX film *Aliens of the Deep,* in which the film crew follows in one submersible the voyage of another crew exploring the bottom of the ocean in another transparent submersible, one of the young women explorers says, "It's like returning from a visit to the future." As Eugene A. Cernan, the last man to walk on the moon, put it, "You can't return home without feeling that difference. But you do come back to reality very quickly. You try to share and relate your feelings to others, but you can't take a billion people back with you. It's almost as if you have come back from the future."

The moon pulls

at the water, lifting up a blue fold in the ocean that may travel a mile or so before crashing into the shore. A wave's matter is continually rearranging—spilling over and into itself to form itself anew. Although we can follow the wave, watch its liquid weave, it consists of different particles two hundred meters out to sea than it does when it laps upon the shore.

Nor are the particles of your body the same. As you eat and excrete, they are continuously replaced even as, and in order for, you to remain physically consistent despite the passing of time. You may have in your body atoms that once grew in the tree of Buddha, soiled the clothes of Jesus, or reflected moonlight in the eye of Picasso. As with the continuous recycling of water in the ocean to make waves, the pool of chemical elements from which we are made is finite. Matter, especially living matter, cycles. With each breath you take, each bead of sweat that evaporates from your skin, or each iron-containing raisin scone that you consume, you replace chemical constituents in your body. In no two consecutive moments are you or any other life-form composed of exactly the same particles.

But life doesn't only use water in its self-organization. Instead it straddles the line between liquid and solid, specializing in colloidal, jelly-like proteins and quivering liquids. Over time, life has gravitated to still-harder, more durable materials—armor and armaments to protect the soft and vulnerable essential parts. For instance, organisms use carbon, phosphorus, sulfur, and strontium in the formation of skeletons. In making itself, marine life uses hard parts. Together organisms such as diatoms and red algae, seals and clams deplete the environment of silica and carbonate, phosphate, iodine, and fluorine. Life concentrates and redistributes, moving components from undersaturated solutions. Some chemicals, like those in the spicules of glass sponges and microscopic radiolaria, are hardened and packaged around the soft tissues of life. Some life-forms drain the ocean of silica to make fiery opal; others sequester sulfates into gypsum or form crystals of barite in their bodies. Living organisms concentrate massive quantities of cobalt, nickel, copper, zinc, molybdenum, chromium, iron, phosphorus, and manganese.

For eons the whole surface of our planet has been disturbed by and conformed to the spreading of a slow "wave," more subtle than a tsunami to be sure, yet judging by the movement of the Earth's surface, far more powerful. An example of such geological movement includes several islands off the coast of Peru, which are formed by hardened droppings of seabirds. As organisms grow and are crowded together on the surface of the Earth, this matter-moving wave becomes stronger. Soviet geochemist Andrei Vitalyevich Lapo writes, "Living matter is a specific kind of rock. . . . An ancient and, at the same time, an eternally young rock. A rock which creates itself and destroys itself to originate again in new generations in the innumerable forms constituting it. The Phoenix of ancient legends . . ." The animate world of scarabs and crabs, of plants and ants, grows from and returns to the "ground," which all organisms continually draw against to construct themselves. The inexhaustibility of the Earth's mineral resources owes to an original cycling economy in which all organisms participate. The Earth is so rich and resourceful, we may forget that our identity is on loan. The very elements of our bodies are owed, not owned. Our debt is to the biosphere. When we die, our accounts are settled as bodies in the soil or ashes in the air, and the materials that made us wend their way into new forms.

The greater our awareness becomes of our surroundings, the more difficult it is to say where life ends and the environment begins. In geochemistry, biogenic matter is made by life but is not in itself alive; this includes things such as coal, pearls, excrement, limestone, honey, milk, and spiderwebs. This term *biogenic matter* alerts us to the difficulty of defining life. This "inanimate" book, for example, is northern conifer plant detritus, from a northern conifer forest. At the same time, it is an industrial product, a human artifact. Although superficially we see it as a mere mass of paper and ink, a deeper view reveals the book to have links beyond the human sphere to global life. Like the redwood tree, with all its deadwood, this book is technically and literally dead. Yet eventually these pages will decompose and be churned by microorganisms or burned by fire. This book, too, will have returned to—if it isn't already part of—the life of Earth.

On Earth, what is alive and what is not? How is life altered to become nonlife, and vice versa? To examine this quandary, let us follow the course of an atom of carbon on an imaginary trip through the biosphere. Coming into your body from, say, an avocado you ate, the carbon atom may already have been spewed forth with smoke and gas from the terrestrial interior through the mouths of volcanoes some thirty times. Belched into the atmosphere millions of years before you were born, it combined with two oxygen atoms to become carbon dioxide, blowing about in the high winds of the troposphere for thousands of years before settling down into the ground to be sucked up by some orange-capped mushroom, say. Eventually the fungus died and an insect piddling about the timber at the forest floor ate our atom. Rains came and the carbon atom moved into a stream and then a river, wending its way into the ocean. Sinking along with the chalky shell of a photosynthetic microbe into the ocean abyss, the atom stuck to the side of a submerged boulder. There it became limestone, joining the microscopic skeletal carbon that formed part of some white chalk cliffs. Even then the atom was not completely cut off from the biosphere's circulation. One day the tectonic plates moved, slamming into each other as magma percolated inside and beneath them, shifting masses of Earth until the atom became exposed to erosion again as part of the land. The same carbon atom burped up through a volcano and into the sky and was sucked down by an avocado tree you eventually consumed in an avocado-and-tomato sandwich. As part of a carbohydrate molecule, the carbon atom helped give you the energy you needed to finish reading this sentence.

Life reigns with a Sphinx-like wholeness that is never exhausted but simply takes further sustenance from all efforts at analysis; categorizing and examining this fullness succeeds finally only in perpetuating life's mystery. The ancient Greek philosopher Heraclitus, known for his view that "being is ever becoming," compared the force that moves the world to a willful child who builds castles of stone and sand only to destroy them and start again. Heraclitus also said that a person can never step in the same river twice, suggesting that the river as a thing must be the water that moves through it, and this water soon goes out to sea. The real river, then, must be not a thing so much as a process, one that maintains an identity despite an astounding turnover of matter. Indeed, a river is not only the water but the water's flow, as well.

Even the English word *being,* perhaps the most static abstract conception of which we are capable, is not static. Being is not only a "thing" but also something that happens.

All of us share this flowing, wave-like nature. Each cell in your body maintains its identity even though it incorporates more than a billion different constituent molecules during its short life span. By extrapolating from tracking the turnover of radioactive atoms, it has been calculated that seven years from now virtually every atom in your body will be different. And since the calcium phosphate of your bones is dead, the bones turn over more slowly; they are less actively metabolized. Although the estimated loss of about a million cells per second sounds damaging, it isn't considering that the human body consists of an estimated forty to fifty trillion cells; on average, then, only one out of forty million of these forty to fifty trillion dies in any given second. Researchers at the Medical Nobel Institute in Stockholm, Sweden hit upon the innovative technique of tracking the radioactive carbon 14 isotope to estimate cell turnover rates. This isotope became much more prevalent in the atmosphere during active nuclear testing, then precipitously dropped about 1963 after the cessation of such testing. Because the radioactive carbon was taken into plants as carbon dioxide, it entered the food chain—including human bodies, where its relative presence can be used to determine the age of various kinds of cell tissues. While skin cells have life spans of only five days, the average age of intestinal tissue is about 11 years, and skeletal muscle tissue 15.1 years; brain tissue tends to be older, with the cerebellum's gray matter only 2.9 years younger than the person, and your occipital cortex gray matter on average about 10.9 years younger than you are. The scientists used genomic DNA to measure tissue age because, not being exchanged after a cell undergoes its most recent cell division, it has the lowest turnover rate of all the molecules that make up the cell. The levels of carbon 14 in the DNA, reflecting the levels in the atmosphere, tell the scientist when the cell was born. The Stockholm work tells us that cells are made from elements outside the body and that the upper limit for cell tissues is usually no more than eleven years. Even the relatively older tissues of the brain continuously take in carbohydrates and lipids and produce new proteins and RNA molecules. While precision in estimating atomic turnover rates remain elusive, it is safe to say that the body is not a stable entity but a metastable vortex, a whirlpool of flesh, blood, and bone, mucus, sweat, and tears, an agglomeration of ever-changing atoms and cells whose continuous turnover is required to create the appearance of stability.

The whole blue-and-white-flecked Earth maintains itself by turning over its components. Ocean carbonate particles, limestone rocks, and animal bodies are continually moving into the soil, with the result that in less than a decade all the carbon dioxide in our atmosphere will have been replaced. In fifteen years the forests in the world will still contain many of the same trees, but each tree will contain many new atoms.

It takes five to six million years for the mass of the world's oceanic reservoir to decompose through the photosynthesizing activities of plants, algae, and cyanobacteria. Hydrogen enters the dynamic bodies of these sun-loving organisms while their oxygen is released into the water and air. The hydrogen and oxygen of water become the organic compounds of living things and the carbon dioxide of the Earth's atmosphere. Addressing the primal blue substance, Antoine de Saint-Exupéry wrote: "Water, you have neither taste, nor color or odor; they delight in you without knowing what you are. One cannot say that you are necessary for life; you are life itself." What we begin to see is that the elements partaking of life exist not only within but also outside bodies as we have traditionally considered them.

In Northeast Spain

at Figueras, there is a museum devoted to Salvador Dalí. Wandering through the corridor on the ground floor of this museum, one gets a feeling for some of the surrealist's love of ambiguity. One giant canvas viewed close up seems to be an abstract painting of colors; looked at through a specially provided scope and from an angle, however, it resolves into an accurate portrait of Abraham Lincoln with beard and stovepipe hat. Still stranger are a series of "sculptures" consisting only of ordinary rocks Dalí has grouped together and arranged at eye level. The rocks are arranged in a pattern to look like the bodies of reclining women: The women at the rocky beach are the beach.

Such works recall the natural ambiguity of beings that are alive. We breathe, eat, walk, talk, write, and think. Yet we are made like Dalí's women-rocks of objects found at the surface of the Earth. The abundance of elements in our bodies mirrors the abundance of elements in the universe at large. What Heidegger would call our "thingly nature" flashes out toward consciousness. Life is a material. It is a mineral, Vernadsky's geological phenomenon. In fact, the nature of matter comes into its own—is seen most clearly—when that matter is arranged as life. Graphite is never more graphite, wood never more wood than when I bear down on this pencil to inscribe these words. The calcium phosphate of human bones is never more so than when it moves inside the fingers leafing through these pages. Even language, signs, the loftiest poetic sentiments and metaphysical flights of fancy, are indissociably physical: stone shavings, slivers of charcoal-marked white bark, electrons coursing as pixels across the laptop monitors of silicon computers. Humanity's conceit that it alone has an access to a privileged realm it calls spiritual may be as unlikely as the disproven notion that we or life generally owes its special attributes to a unique chemical constitution. Rather than lament our material ordinariness, perhaps we should learn to revel in the possibility that spirituality and mind—innerness—are more widely dispersed throughout the natural world. But as we all know, although personal experience is the primary form of data, we do not have direct access to the experience of others; thus, because intentions other than our own are not directly accessible to us, we must base our surmises on others' behavior, their "thingly" activities. This is why the Greeks distinguished between the *Koinos Kosmos* (the shared universe) and the *Idios Kosmos* (the private universe). Science is the realm of the shared universe, the material world, visible phenomena that can be verified among multiple parties. To what extent nonhuman beings have private worlds is not known, but the extent of their machinations and manipulations of materials suggests the possibility of bizarre forms of interiority alien to human consciousness but not necessarily to consciousness itself.

Marine creatures known as chitons (or sea cradles), a form of mollusk, have rows of red and black teeth to scrape algae off underwater coral. The reddish teeth are made of the inorganic iron mineral form ferrous hydrate; the black ones are of magnetite. These teeth are as metallic and as "dead" as a set of nail clippers. The radula, or scraping apparatus capped with tiny iron teeth, of chitons are harder than the chalky reefs they scrape while eating. The rocks, at least some of them, are alive.

In biologically controlled mineralization, an organism's body determines the structure of a mineral or crystal. We see this in the production of the black teeth inside the chiton's body from precursors that thicken and harden. Unlike igneous and metamorphic rocks, produced by the great pressures and high temperatures found inside the Earth, organisms can make minerals at room temperature under just one atmosphere of pressure. Chitons make not only iron but also calcium, such as crystalline dolomite, a fluorinated form of calcium phosphate, the same material that makes up our teeth and bones. And chitons do so throughout the biosphere, from arctic environments to tropical waters, from the one atmosphere of pressure found at the Earth's surface to the more than one hundred atmospheres prevailing in the ocean depths.

Magnetite formed by chitons was once thought to be a unique example of biological mineral processing. But now we know such processing does not occur only with chitons. All organisms have a "thingly nature." Human teeth, for example, are converted toxic waste dumps: Evolutionarily, my teeth derive from the need of marine cells to dump calcium waste outside their cell membranes. Calcium is a mineral that will wreak havoc on normal cellular metabolism. Trucking this hazardous waste across cell lines in ancestral colonies of marine cells may well be the basis of all present-day shell and bone making.

Science fiction has long been fascinated with the tipping point at which sufficiently complex technological objects—computers, robots, and their ilk—become conscious. A common fantasy is that life need not be carbon-based but could be based on silicon or some other element. In a sense, with the evolution of high-speed computers, this has already happened or is on the verge of doing so. We have no way to know if a sufficiently complex machine will attain "inwardness."

In Philip K. Dick's *Do Androids Dream of Electric Sheep?* (the basis for the science-fiction film *Blade Runner*) the Earth, projected not-so-distantly into the future, has become a radioactive wasteland. A nuclear war has driven virtually all animals to extinction, and necessitated the wearing of lead codpieces to protect the human germ line in the minority of those not turned into "specials." These include radioactive rejects such as Dick's Jack Isodore, a repairman who cannot distinguish electronic animals from real ones, and who desires social acceptance from renegade robots. The main, tough male character is Rick Deckard, a married bounty hunter in love with Rachel Rosen, one of the latest line of artificially made humans who nonetheless likes single-malt scotch whiskey and a romantic fling. Described as having attractive small "Irish" buttocks and large "grown up-woman" eyes, she at first appears to be the daughter of the head of the Rosen Association, a rich corporation at the forefront of extraterrestrial android manufacture. The Voigt-Kampff test, however—a sort of polygraph that asks pointed questions trying to stir human-style empathy, such as what would one think about a couch made of human hide—reveals that she is a Nexus 6, the most advanced type of organic robot. With his bounty money Deckard replaces his eponymous electric sheep—real animals are a vanishing commodity in the world of tomorrow—with a very expensive but real live goat, bought from the Rosen Association. This is a huge status symbol. Whether or not computers can experience human emotions is an open question, as is the extent to which we are really free or determined by causes we don't perceive. Dick's real theme is as much empathy as technology. The book opens with a marital dispute over a Penfield Mood Organ, a device that can program emotions, even rather exotic ones such as 481, "awareness of the manifold possibilities of the future," and 888, "the desire to watch TV, no matter what's on it." If Earth's carbon-based life-forms are natural collections of hydrogen and heavier elements such as oxygen, sulfur, and iron synthesized inside stars—which science agrees they are—what separates us from sensitive machines? Or from exquisitely sensitive "lower" organisms with mineral components? Dick's detective Deckard is on a mission to kill that which he loves—a difficult proposition, and one that all life in principle faces. The woman, the goat, the sexy android, his wife, the escaped androids, and even a spider with its legs being torn off all provide the opportunity for reflection on the status of life's common unity, and its ability to recognize itself in itself. Near the book's end Deckard finds a toad—it may be the last one on Earth—and brings it home to his wife, who discovers it is electric.

The robotic paramour, Rachel Rosen, is the killable other, identified as such by signs so subtle, we might be guilty of displaying them ourselves. By identifying the trait that separates the android organic machine from the "real" human as the capacity to empathize, Dick forces us to accept, among other things, the possibility of a future with others who are simultaneously synthetic and real.

But what of the complex mechanical contraptions that, like us, are grown, reproduced, and evolved rather than made? What of the sometimes fantastically complex matter and energy transformers with mineral components that we dismiss as lower life-forms? Although not humanoid in appearance, they withstand scrutiny as possible "organic robots" that may, to varying extents, have inner sensations of experience and encounter with the real, material world.

Calcium minerals (such as calcite, aragonite, carbonate, phosphate, halite, gypsum, and so forth) are the dominant media used in biomineralization. Opal, a semiprecious type of silica known for its iridescent play of colors, comes next. And the magnetic mineral magnetite, far from being confined to the caps of chiton teeth, has been found inside the cells of bacteria, in swimming forms of algae, and in minute quantities in the brains of migratory fish, birds, sea turtles, and honeybees. In many of these species it may act as a compass, serving to orient organisms toward magnetic north. While there are only a limited number of materials on the Earth's surface, the uses to which they can be put seem almost infinite. The segregation and distillation by life of minerals does not violate the second law of thermodynamics, the universal tendency toward atomic chaos. The energized system of life reuses its materials as it produces wastes. The complexity we see at life's surface is compensated by the using-up of energy and the production of heat on a global scale. Indeed, the most complex ecosystems are measurably the coolest when readings are taken from infrared thermometers in low-flying planes or from satellites in space. Life's ability to keep itself cool in the face of a blazing sun is an index of its efficiency at using energy, which it does in the process of maintaining and making more of itself, of growing. And the perception, interiority, and complexity that life to varying degrees displays allows it to tap into the mineral constituents and bionutrients it requires for its own ceaseless growth, reproduction, and evolution. The incessant work exchanging materials at Earth's surface is a natural phenomenon in full accord with thermodynamics' second law. Life reshapes its environment, recycles materials, and disperses energy in accord with the second law. By maintaining themselves metabolically, by growing, reproducing, and evolving, organisms allow energy to disperse more effectively than would be the case without them. Despite creationist claims to the contrary, the sensitive, mineral-manipulating technology of the organism is perfectly natural from the viewpoint of energy science.

The worldwide siphoning of silica by subvisible marine diatoms, and the equally massive removal of calcium carbonate by other microbes, anatomically and physiologically help change the Earth from a mere planet into something more like a body. And as we know, there is nothing artists love to copy more than the body. Aristotle himself recommended the organization of the organism as a model for artists creating works.

Planetary biomineralization is not only massive but subtle. Opossum shrimp use fluorite to avoid the light. These shrimp produce near-perfect needle-shaped crystals of fluorite, which is unlike the fluorite produced by magma inside the Earth. And unlike the organisms that produce minerals from environmentally abundant materials, these shrimp create their crystals from environmentally scarce resources: They are picky. Phosphorus, an absolutely critical compound for the manufacture of intracellular biological materials, including DNA and RNA, is scarce on the surface of the earth: Seemingly simple organisms exert, at least at the material level, taste or selection, choosing fine or rare elements needed to make their subtle energy-transforming forms.

Beautifully symmetrical marine microbes known as radiolarians deplete the oceans of amorphous silica and strontium to produce their ornate skeletons. A shrub in New Zealand has reportedly been found whose dried leaves contain up to 1 percent nickel—a greater percentage than some mineral sources currently being mined. The concentration of vanadium in marine animals known as ascidians rivals the concentration of iron in our own, and yet vanadium is much rarer. These details from the natural world remind me of some of the transformations painted by Belgian surrealist René Magritte. Like the brown wine bottle that becomes a carrot in the painting titled *The Explanation,* or his depictions of pointy leaves broadening into the shape of green-beaked birds, life forms itself from the palette of whatever is available: Flowers grow from soil; fungi grow from trees. A petrified forest intrigues us precisely because wood becomes rock. In art and nature there is a difference between what an object is made of and what an object is. "Don't you see," wrote Giordano Bruno, "that which was seed will get green herb and herb will turn into ear and ear into bread? Bread will turn into nutrient liquid, which produces blood, from blood semen, embryo, men, corpse, Earth, rock and mineral and thus matter will change its form ever and ever and is capable of taking any natural form." Distinct materials fill the same forms. In life the forms have a function, to expend energy. This is the same function that other nongenetic but highly organized complex thermodynamic systems have. The forms of art may be "merely" decorative, but the economy of great works of art seems to mirror that of nature.

In producing themselves, such beings as the ascidian tunicates and opossum shrimp have opted for the rarest of materials. With them and their "expensive tastes," living sculpture has been raised to a fine art.

The development of modern art even parallels a trend in biomineralization. The nautilus is a "living fossil" related to the squid and octopus. Its ancestors were once much more prevalent in the world's oceans. The nautilus has a powerful aragonite "beak" capable of crushing bones. Its shell is also aragonite, while its balancing organ is formed of calcite; in addition, it has normal "kidney stones" formed of shelly phosphate minerals.

The nautilus and its descendants illustrate biomineralization as an evolutionary work in progress. The balancing organs in this organism are formed from loose assemblages of crystals in a "pinpoint mineralization." But over evolutionary time this loose polycrystalline network has been sculpted together. In the balancing organs of more recently evolved descendant cephalopods such as the octopus and squid, the scattered pinpoint array of crystals has coalesced. All nautilus-type organisms have kidney stones, but in some species the kidney has lost its function. Nonetheless, because the need to make shells is so strong, the kidney stones have been retained. They continue to serve as reservoirs storing the raw skeletal materials calcium and phosphate within the organism.

Like romantic knights of the Middle Ages, in the ancient oceans marine creatures evolved protective armors. They formed poking appendages and rock-cracking jaws to break down coarse or protected foods. But eventually speed substituted for armory. Coy survival ploys such as the ink squirted by the squid and the intelligence of the octopus came into play. Today the squid and the octopus lack shells. They do not biomineralize like their nautilus ancestors. Yet their responsive bodies, and ours, still use calcium to think and react. Calcium crucially mediates in the cell-to-cell interactions thought to be the neurological basis for reflex and reflexive thought. Calcium is lethal to cells in a free ionic state, and although calcium ions are ten thousand times more prevalent in the oceans than is the poison cyanide, calcium itself has been incorporated into the very marrow of life, into its skeletons and shells, and into physiological processes ranging from blood clotting to thinking. When Dalí lay down on a Spanish beach to tease forth the image of a woman, when he sketched a nude reclining in a seabed with nothing but pigment and chalk, his brain activity was part of the planetary use of calcium as a medium in a masterful, if unconscious, three-dimensional composition.

Water—that aquatic acrobat, that master of Ovidian metamorphosis, curls and sinks in heat conduits. This taker of forms, of ice and snow and rain and rivers and water, scientists concur, is a key to life on Earth, if not life elsewhere in the solar system or the universe. It was once considered to be one of four essential "elements" making up the cosmos: water, fire, air, and earth. At a time when philosophers and scientists were one and the same, Thales of Miletus, the founder of Greek physical philosophy, described by some as the first scientist, postulated water to be the primordial stuff of being. Although he believed, according to Aristotle (in *De Anima*), that all things were "full of gods," he did not think the gods were responsible for the origin of life. A statesman and visitor to Egypt, Thales is considered the earliest materialistic Greek philosopher because he discarded mythical accounts of origins. Anaximander, another Greek philosopher, criticized Thales' idea that everything originates in water: Anaximander pointed out that water destroys that great source of many other things, fire. And Anaximedes, another Greek, criticized Thales' naturalistic postulate, suggesting that perhaps air was the prime motivator of earthly phenomena.

Under water

we might also classify the wonderful plants that, with the recycling, nutrient-gathering fungi grew forth as the grass carpets and colorful jungles supplying our hominid ancestors with their settings and sweets, their mental stimulation, physical sustenance, wood-fed fires, and woodland homes. Our soaps, medicines, plastics, our dyes and candy and gum and lemons and limes and pineapples and linen sails and hallucinogenic alkaloids with mystic messages of the secret structure of the universe—sometimes in the comic form of the ingested plant itself—our cheesy mystery novels, sacred texts, important documents, paper money, pulp science fiction, and this very page are part of what plants do for us in return. Although silent, the botanical gods of the rain forest and savanna play tricks on our minds, "getting us" to eat their fleshy fruit in order to tend their bodies and spread their seed. As proto-human primates discarded the pits of the coveted fruits whose bright colors were prototypes of the seemingly unstoppable barrage of modern advertising, they accomplished, without ever giving up the feelings of mastery, pretty much exactly what the plants required. They are kinder, gentler versions of the pods from outer space that take over human bodies and modify human behavior to aid in their own propagation in the film *Invasion of the Body Snatchers.* And they are real. Without making any visible effort they get us, green zombies that we are, to do pretty much the same thing we do for ourselves when we go around chasing money, making dinner plans, and taking care of children. They promote their own kind, one version among several major types of the cyclical chemical organization, genetically riveted and thermodynamically active, that inhabits this wild Earth.

Which leads us by and by to the importance of trees from a global warming perspective. Water is a major greenhouse gas, but its effects as a reflective surface may be even more important. Complex thermodynamic systems tend to maintain their complexity while taking advantage of local energy streams, which they naturally sap. The existence of a complex thermodynamic system, including a living one, saps the environment of energy, degrading gradients faster than if the cycling system were not there in its intriguing and ornate beauty.

A gradient is a measurable difference across a distance, for example a difference in pressure or temperature. Revealingly, the natural appearance of swirling systems such as, say, a tornado in the atmosphere rectifies a local barometric pressure difference, producing entropy, mostly as heat in the process.

Although carbon dioxide is generally credited with causing global warming—by letting in light, which it then traps as heat (like a greenhouse)—as or more important may be water. Eric Schneider, a geologist who for more than twenty years (and still going) turned his attention to the thermodynamics of ecosystems, has postulated that advanced ecosystems such as rain forests are more thermodynamically complex, emitting more entropy and degrading gradients more effectively, than farmland or suburbia. Indeed, it turns out that rain forests are the most effective gradient reducers on the planet. Satellite thermal readings of Amazon and Borneo rain forests at the equator in the height of summer are cooler than Siberia in the middle of winter. The cooling seems counterintuitive until you realize that the satellites are measuring not equatorial heat but what life does with the solar energy. Rain is an integral part of the jungle ecosystem, as the term *rain forest* suggests. The temperature measurements are far lower than might be expected because the ecosystem extends beyond the gas-exchanging boundaries of organisms to the gases themselves. Measured from space the lush, would-be hot ecosystems are cool for a reason.

It's because of the clouds.

AIR

Lighter than the body
Better than the soul . . .
—JOSEPH BRODSKY, "THE CLOUDS"

For years
people have wondered about life on Mars. Boston astronomer Percival Lowell, whose body lies at the base of the Flagstaff, Arizona, telescope on Mars Hill, "discovered" irrigating "canals" on the red planet. On Halloween 1938 Orson Welles made a convincing radio broadcast simulating an invasion of New York City by Martians, and many people, ignoring the disclaimers, were frightened out of their wits. When the *Viking* Lander plunked down, however, it found no flowing waters or agricultural fields but only a dry, ruddy desert as far as the camera eye could see. The irony is that the camera eye itself represents the arrival of the life for which we are searching. In my mind's eye I picture lovers lolling by the moist bed of Martian canals, a Mars covered with a latticework of waterways irrigating lemon trees and orange groves—a colorful paradise not found but created. Perhaps, as the philanthropists on Huxley.net dream, they will have combined genetics, neurobiology, pharmacology, and nanotechnology to create a world free from pain. Among the easy livers will walk hand in hand the postmordial Adam and Eve. And if the future becomes a template for the past, the mother of all to come will whisper sweet somethings to her soon-to-be-risen lover, and the whole accursed sideshow will start off on the right side of the bed this time.

I mention this science-fiction fantasy of technological salvation not only because it is a shining example of hubris, but because it exemplifies the endless possible perspectives that, thank God, have not been monopolized by either religion or science. The future scenario of a hypothetical good Eve who whispers in her Martian garden segues to that fraught topic that, like water, is so present as to be invisible but that, unlike water, *is* invisible: the air. Especially fresh air: in the woods where you might feel the wind on your face, watch the clouds overhead, and hear the hoots of an owl as you walk on soil aerated by ants whose collective weight is greater than that of six-plus-billion humans.

Out of thin air is just a metaphor for our ignorance of not knowing where something comes from. But the air is not nothing. It is something.

Air itself is the molecular medium for land animals. The preferred medium of communication for the olfactory set, it is also the sine qua non of speech, which requires something through which to send sound waves. One can't talk face-to-face without an atmosphere. When aquanauts submerse, they carry their air in their submersibles. When deep-sea divers dive deep, they carry it on their back. When the astronauts landed on the moon, they had their air tanks with them. We take it for granted because it is everywhere. Only when leaves swirl, or a telephone book mysteriously turns its own page, do we realize it's there.

Inspiration—
which hides the word *spirit,* as do the kindred *respiration, aspiration,* and *expiration*—comes from the Latin word for "breathing," *spiritus.* Its etymology, according to Merriam-Webster's, is "Middle English, from Anglo-French or Latin; Anglo-French, *espirit, spirit,* from Latin *spiritus,* literally, breath, from *spirare* to blow, breathe."

The first of Merriam's thirteen entries is the scientifically suspect but culturally embedded notion that spirit is "an animating . . . principle held to give life to physical organisms." According to scholar Hans Jonas, this prescientific hypothesis, at the origin of the notion of gods, came about when people reasoned from the evidence that things could move without wills to the conclusion that wills—animating spirits—could exist separately from bodies. There was thus a wind spirit, a moon spirit, a mountain spirit, and so on—eventually they become gods, and then God. Merriam's second of thirteen definitions is "a supernatural being or essence: as a *capitalized* : **HOLY SPIRIT b** : **SOUL 2a c** : an often malevolent being that is bodiless but can become visible; *specifically* : **GHOST 2 d** : a malevolent being that enters and possesses a human being." We can be in good spirits, in high spirits, or mean-spirited. We can't prove it scientifically but life seems to be spiritual as well as physical.

We are living in the Holocene, when anything is possible. According to Sir James Jeans, the fundamental realizations of modern physics suggest that our normal view of reality is not reliable. It may be that, although we think of ourselves as isolated and alone, we are really aligned on a deeper, spiritual level. Individuals appear as particles but are really indissociable parts of a continuous wave. The ultimate makeup of the cosmos, in which the relationship of parts to wholes is strange indeed, may apply also to us in our ultimate reality as human beings:

> In the particle-picture, which depicts the phenomenal world, each particle and each photon is a distinct individual going its own way. When we pass one stage further towards reality we come to the wave-picture. Photons are no longer independent individuals, but members of a single organization or whole—a beam of light—in which their separate individualities are merged, not merely in the superficial sense in which an an individual is lost in a crowd, but rather as a raindrop is lost in the sea. The same is true of electrons; in the wave-picture these lose their separate individualities and become simply fractions of a continuous current of electrons. In each case, space and time are inhabited by distinct individuals, but when we pass beyond space and time, from the world of phenomena towards reality, individuality is replaced by community.
>
> It seems at least conceivable that what is true of perceived objects may also be true of perceiving minds; just as there are

There are no parts.

> wave-pictures for light and electricity, so there may be a corresponding picture for consciousness. When we view ourselves in space and time, our consciousnesses are obviously the separate individuals of a particle-picture, but when we pass beyond space and time, they may perhaps form ingredients of a single continuous stream of life. As it is with light and electricity, so it may be with life; the phenomena may be individuals carrying on separate existences in space and time, while in the deeper reality beyond space and time we may all be members of one body. In brief, modern physics is not altogether antagonistic to an objective idealism like that of Hegel.

As Alice Walker says in regard to race relations, there are no others. They're all us. Although we seem alone we perhaps belong to an unseen whole, and our individuality may in fact be projection points bubbling up from an unseen underlying multidimensional reality. Be that as it may, the epistemological point is that there are many ways to make spiritual sense of the available reality. That all is not as it seems is clear from University of California neuroscientist Benjamin Libet's experiments. He showed that people, using an external timer, consistently misidentify the moment of decision making. Although will appears to be free, and perception gives the impression that our senses are mere windows onto the outside world, consciousness may be more like the operating system of a computer—a mere interface, dealing in perceptual shortcuts, that hides the intricate inner workings of the brain even as it whittles down the vast data set of raw reality into a manageable subset that we experience as the unadulterated whole. Psychotropic drugs don't necessarily distort reality so much as they reveal it in the raw, exposing to an overwhelmed operating system the writhing richness of the eternal innards. Even free will, as the Libet experiments suggest, may be illusory. Assuming that this is the case, the question arises: *Why do we have the illusion of free will?* When I talked to German chaos mathematician Otto Rössler, coming up to him after he had mumbled through some suggestive overheads in Madrid, he said he had an "Aristotelian solution." By which he meant that an external creator introduced free will into the "chaotic" mix. *Chaos* here does not mean "random": It means, in a model of which science has been long enamored, the algorithmic production of life, the universe, and everything from particles in a void operating deterministically according to universal laws. For Rössler free will really is free—a little bit of divinity sprinkled into a mechanical universe. If Libet is right, however, the divinity is either sprinkled everywhere or nowhere: Human freedom is a seductive illusion. Determinism reigns supreme, even at the heart of the human soul.

Holder of mildew;
medium of smell and sound; distributor of seeds. Supporter of helicopters; repository of sighs. You inspire me; I respire you.

Trapped in ancient bubbles of ice, you were made when snow fell before our great-great-great-grandparents, preserving what was then. You look like but are not nothing. You are—the singular plural. Your molecules refract sunlight. Your beautiful impurities take the shine off the sun, mute its brilliance into pastoral pastels. You are the olfactory ocean of the canine world, triggering memories older than any words.

The medium we do not see is not water, but air. The atmosphere is the circulatory system of Gaia. Could we be a cosmic organism the size and shape of a planet's surface, no more a biological fiction and no less separate and self-acting than each of us? Are our megalopoli interconnected neuron centers within a developing global tissue, the diffuse brain of the macroorganism whose unconscious, let alone conscious, capabilities dwarf our own?

For life to inhabit a planet, it must make use of that planet's atmosphere, its external gaseous circulatory system. It is this medium through which chemicals cycle on a global scale. On Earth all we say and do occurs within this medium, which may, just may, be the real tissue of a being.

In "Merleau-Ponty and the Voice of the Earth," eco-philosopher David Abram suggests that, like technology, language is not our own but in some sense owned by and owed to the biosphere:

> If language is rooted in perception, then it is never, in actuality, a language of wholly abstract, ideal, or purely mathematical relations, for it is inhabited by all those things, styles and rhythms to which our senses give us access. Indeed, we may begin to discern that this our language has been contributed to, and is still sustained by, many gestures, expressions and sounds besides those of our single species! A language that has its real genesis in the deep world of untamed perception is a language that was born as a call for and a response to a gesturing, sounding, speaking landscape—a world of thunderous rumblings, of chattering brooks, of flapping, flying, screeching things, of roars and sighing winds. . . .

That is why Merleau-Ponty could write, in his unfinished work, that "language is everything, since it is the voice of no one, since it is the very voice of the things, the waves and the forests. . . ." "We may even begin to suspect," says Abram, "that this language we speak is the voice of the living Earth itself, singing through the human form. . . . Logos is realized in us, but is not our property."

In the science-fiction work *The Android's Dream,* by John Scalzi, members of an advanced race communicate with one another by farting precisely in an olfactory semiosphere. The plot begins with diplomatic difficulties when a human, with a rectal translating device, farts out insults in the scent-language of the alien race.

We laugh when we realize, at a gut level, that we don't know. There is a gap in our logic, a double meaning or misunderstanding. Co-discoverer of the DNA molecular structure Francis Crick believed he had found the physical location of free will in the brain. Think of a scientist announcing the discovery of the locus of humor in the cytoplasm of human cells. The very idea that we can locate something as amorphous as the sense of humor is hilarious. When we don't know something, when our attention is brought to our gap in knowledge, or to our base physical needs, we laugh. We blow short gusts of wind into the air.

Born in St. Petersburg
on March 12, 1863 (and therefore, like me, a Pisces), Vladimir Ivanovich Vernadsky wrote that his intellectual awakening took place during his youth on walks with his father (a political economist and statistician) and his father's cousin, the retired military officer Yevgraf Maksimovich Korolenko.

Under the starlit skies, the gray-bearded "Uncle" Korolenko—an atheist familiar with the works of French naturalists Jean Baptiste Pierre Antoine de Monet de Lamarck and Comte de Georges Louis Leclere Buffon, as well as British geologist Sir Charles Lyell and naturalist Charles Darwin—speculated that life existed upon other planets, that thought and matter were inseparable and could give rise to the other, that matter circulated on a worldwide scale through plants and animals, and that the Earth was a living organism. Six years after his mentor Korolenko died, Vernadsky wrote, "It sometimes seems to me that I must work not only for myself, but for him as well, and that not only my life but his life as well, will have been lived in vain if I accomplish nothing." Such ingenuous sentiment, humbly acknowledging his quasi-familial influences, contrasts with Charles Darwin's studied avoidance of his intellectual predecessors, including his own grandfather Erasmus, whose *Zoonomia,* an epic poem that mentions natural selection, most likely could be found in Darwin's father's library.

Although Vernadsky is not so well-known in the West, in the former Soviet Union postage stamps were designed with his visage. As the founder of modern biospheric thinking, he may be said to have done for space what Darwin did for time: expanding our scientific view of organisms.

His early exposure to Korolenko's ideas provided Vernadsky with lifelong inspiration. Vernadsky broadened his studies. He became interested in the effects of solar radiation on life, and of life upon itself and the surface of the Earth. All the while he maintained that "The right of freedom of thought for me is one of the most necessary preconditions for a normal life, and I could never tolerate the lack of it."

Originally a mineralogist and crystallographer, Vernadsky came to biology through his realization that the properties of soil could not be understood without taking into account the enormous influence of the living beings that help form soil, and that soil contains.

Austrian geologist Edward Suess first used the term *biosphere* in 1875 in a slim book on how mountains arise. Suess pointed out, "One thing seems to be foreign on this large celestial body consisting of spheres, namely—organic life. . . . On the surface of continents it is possible to single out a self-contained biosphere." Vernadsky first encountered the term in reading Suess's last work, a multivolume treatise called *The Face of the Earth,* a monumental synthesis of all geology up to that time, composed between 1883 and 1909. Despite this introduction of the word into the scientific literature, it did not catch on at first except among a few intellectuals studying in Paris after World War I. Vernadsky popularized *biosphere,* noting: "[T]he environment in which we live, it is the 'nature' that surrounds us and to which we refer in common parlance." Among those who enthusiastically embraced the term *biosphere,* and explored the implications of its use from the start, were Vernadsky and his contemporaries Edouard Le Roy (philosopher Henri Bergson's successor at the College de France) and Pierre Teilhard de Chardin, the well-known "evolutionary theologian" who was then professor of geology at the Institut Catholique in Paris. The biosphere concept, elaborated by both Vernadsky and Chardin, implies that life is not simply a matter of individual plants and animals but a planetary phenomenon. The related term *noosphere*—referring to the sphere of human influence on the planet—also used by both men, but in very different ways, was first bandied about in Le Roy's books but may have arisen during a meeting of Chardin, Le Roy, and Vernadsky in Paris during the 1920s. Whereas *noosphere* for Chardin was the "human" planetary layer forming "outside and above the biosphere," for Vernadsky the term referred to humanity and technology but as inside and part of the planetary biosphere. Swiss historian Jacques Grinevald explains that the distinct purposes to which the theistic Chardin and the atheistic Vernadsky put these terms have lead to the duality of the term *biosphere* as currently used: On the one hand, *biosphere* refers to the zone in which life exists (which was the way Vernadsky usually employed it); on the other, it means the totality of life itself (as Chardin used it).

Vernadsky came to use the word *biosphere* to mean more than a thin film at the Earth's surface. The study of soil provided a focus for the interaction of life and minerals, for a view in which life was a special kind of mineral, part of a planetary chemical reaction. Vernadsky came to believe that the diversity of life-forms, far from being incidental to the planet's surface, were essential to many of the minerals of the Earth's crust. He was right. The thin layer of living matter made an impact all out of proportion to the scanty space it occupied. He believed that, of all the geological forces acting on the surface of the Earth, living matter was the greatest, and its force increased with time. We can see Vernadsky's "force of life" in the horizontal spreading of life to all parts of the Earth's surface, as well as in its ascent from the oceanic depths to the land and atmosphere beyond. This "force" is really the potential energy stored in life as it cycles matter and grows, including into itself, as a complex system on Earth's surface.

Vernadsky retained the child-like wonder of his walks with Korolenko. Sparked by this youthful insight and keen perception, he continued to be amazed with the ways of nature, even after others felt such ways were well understood and had been adequately explained. We now know that life is an open thermodynamic system whose natural entropy-producing complexity, cycling matter chemically for perhaps four billion years, is ensured through gene replication.

Like many visionaries—including Lovelock, whose startling realizations of the implication of planetary chemical disequilibrium were discovered serendipitously by the attempt to find life on Mars and to think of detecting life on Earth from afar—Vernadsky looked at his subject matter from a broader perspective. He looked at life—properly the subject matter of biology—from a standpoint beyond biology: the vantage point of chemistry and astronomy, crystallography and natural philosophy. To him living matter—he avoided the Russian term for "life"—was an organic mineral with some very unusual properties. It was a mineral, mostly water, whose crystalline regularity and chemical impurities worked to gather the energy to spread across, within, and as our planet's surface. The source for life's planetary spreading, its concentration of the energy forms that manifested in growth, reproduction, and thought, was not even terrestrial. It was solar:

> The biosphere is as much, or even more, the creation of the Sun as it is a manifestation of earthly processes. Ancient religious intuitions which regarded terrestrial creatures, especially human beings as "children of the Sun" were much nearer the truth than those which looked upon them as a mere ephemeral creation, a blind and accidental product of matter and earth-forces. Terrestrial creatures are the fruit of a long and complicated cosmic process and, subject to predetermined laws, form a necessary part of a harmonious cosmic mechanism in which chance does not exist.

Vernadsky pictured life on Earth as a global chemical reaction, a "green burning." Where others saw groups of species and organisms, Vernadsky saw patches and films of atoms collecting and migrating under the influence of the sun. A radiation-trapping and -transforming mineral, living matter spreads and grows, it coalesces and disseminates, it merges and diverges with itself. In his ultramaterialistic perspective, Vernadsky saw through the word *life* to a strange solar-powered mineral. Living matter, capturing and redeploying solar energy, continuously remakes Earth's surface. Mystic as it might sound, Vernadsky's idea that life is not the essence of some abstract quality but rather a physical thing with the qualities of a unique mineral is in fact the antithesis of antiscientific superstition. Indeed, I would argue that the commonplace four-letter word *life* is intrinsically vitalistic. Vitalism is the obsolete superstition that the laws of physics and chemistry do not apply to life because it is made of some special substance, or endowed with some supernatural power. It was once thought that decaying meat spontaneously converted into maggots, that the progeny of an animal was in an infinite number of seeds in the female, that living forms unfolded according to a hidden plan called an entelechy. All of these can now be seen to be misguided versions of the second Merriam-Webster definition of *spirit,* which highlights the old human belief that living matter, including ourselves, is inhabited with a transdimensional visitor, temporarily housed in human form. Such fleshly incarnations are anathema to science. The belief in a separate spirit, though natural, may be a cultural holdover combined with a sort of cognitive illusion. So deep is the desire or natural inclination to believe that those trying to prove the existence of subtle body jumped for joy when corpses were found to be lighter than their corresponding bodies—as if the spirit, with slight mass, had taken off. Of course, what had vanished was the water that evaporates from the body at the time of expiration. Generations earlier, an American movement of devout believers in the possibility of life after death—exacerbated by the secret table tapping and other tricks of spiritualists holding séances to contact the beloved departed—were publically opposed by the escape artist Houdini. I would argue that to the very word *life* clings the residue of a vast array of cultural belief systems identifying the phenomenon with some hidden supernatural principle or power. From this perspective, Vernadsky's jettisoning of the intrinsically vitalistic *life* in favor of the adjectival *living matter* was prescient. There is no way to eradicate the ectoplasmic aftersheen of vitalism from the simple, ordinary, and seemingly philosophically uncontentious term *life.* But scientifically, life is not some *vital* stuff. The word itself, in its implication that it refers to a discrete substance and separate process, is as nonsensical as Molière's scientific joke that sleep is caused by a "dormitive principle."

After lifelong study, Vernadsky reached two major conclusions, which he raised to the status of biospheric laws. He called them biogeochemical principles. The first was that "The biogenic migration of the atoms of chemical elements in the biosphere always tends to its maximum manifestation." This means that, over time, the atoms that form and re-form organisms become busier and broader. More and more atoms, more and more elements become involved in the expanding Earth-life system. The biosphere grows in extent and activity.

The second biogeochemical principle holds that "The evolution of species in the course of geological time, leading to the creation of life-forms that are stable in the biosphere, proceeds in a direction which increases the biogenic migration of atoms in the biosphere." The origin of new species, in other words, has a global biological context. Evolution, we might say, takes place in and as Gaia. And *Gaia,* as merely our name for the vast thermodynamic system of the biosphere, does what all natural complex systems in the universe do: It produces, within constraints, as much entropy, as much atomic chaos, mostly as heat, as possible. There is no more effective way to do that than by growing, unless it is by evolving species, such as ourselves, able to use perception, intelligence, and/or technology to search and deploy new sources of energy. At the same time our global increase of entropy production by increasing the extent of our "advanced" technical civilization has interfered with older, more stable forms of entropy production—such as rain forests and old-growth forests, not to mention greenhouse gas emissions increasing global mean temperatures—and therefore the biosphere's ability to cool itself.

From a Vernadskian perspective, organisms and species are mineralogically to be analyzed as elementary movements and atoms in flux. Our atmosphere contains rapidly moving, chemically anomalous particulates of calcium, carbon, and phosphorus. We call them seagulls. So, too, do a hundred trillion ants in the Amazon add about fifty-five thousand tons of formic acid a year to the environment, generating maybe one-fourth of all the acid rain there. Fungi on rotting trees leak some five million tons of chlorocarbons into the atmosphere a year by disseminating their spores. Life is a form of active, energized rock: Geochemically, a swarm of locusts is the point of transformation of one section of the biosphere into another. It is a flying mountain. In 1890 G. T. Carruthers calculated that the weight of a swarm of locusts—grasshoppers in their swarming phase—documented to have covered the sky above the Red Sea for two days and a night was forty-four million tons. The plague of locusts descended on fertile Mesopotamian fields, devoured everything, then flew to Ethiopia. This quantity of transported mass equals all the zinc and copper mined throughout the nineteenth century.

Just as Lovelock's Gaia hypothesis, which regards Earth's surface as the largest organism in the solar system, forces us to regard the nonliving environment as parts of a great body, so Vernadsky, by resolutely treating life as a chemical and mineral phenomenon, opens the way toward a more integrated planetary perspective. Although Lovelock disparages Vernadsky for presenting only "anecdotal" ideas and not being a true scientist, Vernadsky, like Darwin, was resolute in not excluding humankind from his analysis of the life process. Humans partake of the biospheric principle that over evolutionary time more chemicals become involved in the biospheric circulation. Indeed, in our global market and scientific enterprise we circulate many compounds, from new pharmaceuticals and black-market drugs to synthetic plastics, copper coins, and military munitions (think platinum, uranium, steel). Some of these have been rarely used by other organisms. Garbage disposals, automobiles, and factory exhaust increase the rate of atomic migration at the Earth's surface. Using cyclotrons so expensive they require multinational intergovernmental funding, particles and atoms rarely if ever seen, at least recently, in our corner of the cosmos, have been produced. Life no more violates thermodynamics' second law than human technology violates Vernadsky's first law.

Considering Earth a giant organism and life a mineral phenomenon may seem a contradiction, but both expand our focus, breaking down the academic barrier between biology and geology, which reflects academic interests and specializations, not global operations. Fully 50 percent of the nitrogen atoms in your body were recently in a fertilizer factory. Life is a process. It cycles matter and produces entropy, mostly as heat, to keep itself cool. Keeping cool is not specific to life. It is a thermodynamic property of many complex systems that sustain themselves or are sustained by energy flow.

To me it is revealing that Lovelock's epiphany regarding life's planetary status was the result of realizing that the atmosphere was far from thermodynamic equilibrium. Vernadsky's focus on life as a mineral phenomenon, with global cycling behaviors energized by the sun, is also a thermodynamic perspective. Philosophical habits of thought and linguistic patterns of speech die hard. They still reinforce the ultimately insupportable notion that life is a thing apart from the universe. Life is utterly dependent upon, and only conceivable within the context of, its external materials and energy-rich surroundings.

The biosphere is like a redwood tree: Although only the outer part is actively growing, we think of the entire tree—including its "dead" wood and bark—as alive. The chalky cliffs, the organic-rich oceans, and the allusive clouds, far from being the painted scenes behind the live actors in an inert environment, are part of the extensive body of a unified process, active and growing. Lovelock describes the term *biosphere* as a "vague, imprecise word that acknowledges the power of life on Earth without surrendering human sovereignty." I understand Lovelock's objection. However, *Gaia* has been for some too tainted by mythology and the putative horrors of pre-religious, pre-scientific, and animistic—pagan and feminist—worldview to be accepted. Ironically, religious Teilhard de Chardin's use of the term *biosphere* as the totality of life more closely overlaps *Gaia* as used by scientists today. However, neither Vernadsky nor Teilhard de Chardin had an understanding of the thermodynamically based feedback of the global living system. Vernadsky focused on life as a mineralogical phenomenon, Teilhard on evolution as a physical manifestation of spirit heading toward greater complexity. In the end, as in the beginning, God would be. He would know what he was through what had evolved, which was also created. Evolution was not merely random; it had a direction. It was going somewhere.

In contrast to our view of ourselves as the prima donna, the most important actor on the evolutionary stage, in fact we may be involved in an exponential evolution that goes far beyond us. If we regard the biosphere as a giant body, we can make the case that it is indeed developing something much like a planetary neural system. What Vernadsky and Teilhard in their different ways call the noosphere can also be termed an exobrain. Just as insects have exoskeletons, global humanity with its technology and telecommunications has an exobrain. I translate the term from the Spanish original *(exocerebro)* contained in a reference to R. Batra's *La conciencia y el exocerebro* in a paper on art, science, and politics by Madrid-based cultural historians Karin Ohlenschläger and Luis Rico. Perhaps, Batra writes, "humans' cerebral tissues have looked beyond the weak skull hiding them and found an artificial exobrain, exposed to the elements, that gives them the support of a solid symbolic structure." Lovelock talks of us being the neural tissue of Gaia. Others have developed the idea of a global brain. The US mathematician, chemist, ecologist, and general free thinker Alfred Lotka (1880–1949) wondered if thought, able to transform matter, was not itself a form of energy. The vast complexity of Earth's environment, like the ultimate nature of the universe, eludes simple human minds; perhaps our cognitive lack, our finally coming up short, in trying to find the grand design or story of the universe, is a structural necessity. We are only a minuscule part of the vast colossus we examine. To even begin to comprehend that which contains us, and so much more, we caricature, simplify, and distort. We dangerously extrapolate from our own experience, see ourselves as playing a central role, and as often as not make serious additional mischief based on our own misconceptions.

A classic example is what ecologist Garret Hardin called "the tragedy of the commons," based on the degradation of common grazing areas in England: Individual farmers, feeding without allegiance to the larger whole, ruined the grazing land for all. This phenomenon is general, as seen in the driving to extinction by early humans of many presumably delicious food species—not to mention global air pollution, overfishing of the oceans, and possible irrevocable climate-based destruction of the global commons of ocean and atmosphere. There are no necessary natural limits to growth, but growth entails pollution, generates additional pain, and leads to more death as organisms vie for limited resources. If not morally or chemically restrained into a higher collective, populations will exhaust their food sources. That gene-trading bacteria, more complex amoeba-like cells (eukaryotes), and other beings have come together to make new sorts of cells and multicelled organisms is not surprising. We are surrounded by life, but as humans we belabor under the cultural illusion that we are largely in control.

Teilhard de Chardin spoke of the Omega Point, the hypothetical culmination of earthly and human evolution in the far distant future as all people converge into a single Christ consciousness. Such views, flattering to humanity's sense of its own importance, have been resuscitated in recent times. Might it be possible to create paradise, or God, or immortality in the future via technology—realizing in the future religious stories of idyllic perfection in the past? Is it possible that a Supreme Being, even if he is an anthropocentric fantasy concocted by priests in the days of old, may arise as a logical culmination of the evolution of which we ourselves are part? Physicist Frank Tipler finds hope for human immortality in the thermodynamic facts that energy changes form but is never created or destroyed. The "singularity" of science-fiction writer Vernor Vinge and technologist Ray Kurzweil is also a move in this direction: Technological progress, especially an "intelligence explosion" in computers (which may induce them to "awaken"), will inaugurate evolution's dramatic next step. The term *singularity* in this futuristic sense arose when mathematician John von Neumann had a conversation with colleague Stanislaw Ulam that, "centered on the ever-accelerating progress of technology and changes in the mode of human life, which gives the appearance of approaching some essential singularity in the history of the race beyond which human affairs, as we know them, could not continue."

Again science fiction illustrates in story what futurists describe in theory. In a tale by Frederic Brown, the world pools its scientific resources to build a giant computer gathering and synthesizing all human wisdom. Once it is operational, someone tremblingly types in the question, "Is there a God?" The machine answers at once: "Yes, now there is. . . ."

Perhaps the singularity has already come without us lowly humans recognizing it. In fact, this third blue planet from the sun may consist of an advanced life-form that perfected pollution-free waste management, parallel-shape-based molecular computing, and nanotechnological materials processing billions of years ago. In the end man's contribution may be a limited part, a bit player, a transitional form helping to develop means of celestial technological dissipation of the Gaian life-form. The fate of this organism may be to reproduce itself prolifically throughout the solar system and perhaps beyond it. Even if we stoically refuse to consider it either alive or endowed with the sort of spirit or mind we detect in ourselves, we can glean that its development, toward increasing complexity, carrying us with it, is not exactly random.

As in the lines from Gnarls Barkley's song "Crazy," "You really think you're in control/Well, I think you're crazy," we humans think we are in control and, to the extent that we do, we may well be crazy. Life is too rich and nonhuman to cling to the emotional security blanket of the delusion that we are in planetary control. A molecular biology study using genetic assays rather than counting microbes that can survive in culture found that marine water contains some five to ten million kinds of microscopic organisms, rather than the five hundred thousand kinds previously measured. Part of a ten-year Census of Marine Life begun in 2000 at the Marine Biological Laboratory in Woods Hole, the new, improved practice of quantifying species by genetic snippets has been likened to the improvements of the Hubble Space Telescope over an ordinary telescope. Norman Pace, a molecular biologist at the University of Colorado, says our old understanding of microbiological diversity was statistically skewed due to our information-gathering methods—as if our knowledge of animals were based entirely on a visit to a zoo.

Earth's surface chemistry and reactivity is a mass phenomenon created by the pooled metabolism of the diverse living creatures it contains. These creatures—soil and cyanobacteria—are capable of metabolic feats, such as taking nitrogen from the air at room temperature and producing water as a waste product from the use of hydrogen as a clean fuel, that still elude the best scientific minds and laboratories. Researchers at Boston University found more than 115 different types of desiccation-resistant microorganisms present in seaside muck; the organisms, none of them plants or animals, that grew upon being wetted in the laboratory were not common in the field. Biologist Margaret J. McFall-Ngai, a researcher at the Kewalo Marine Laboratory in Hawaii and international expert on animal–bacterial symbioses, speculates that the origin of the immune system is connected to our marine ancestors' need, in oceans with up to a hundred million microorganisms per liter, to welcome in some but not other microbes. She recalls stumping a roomful of doctors with the question: *What might be the effect of our ancestors being surrounded by so many "germs"?* Her answer, in short, is that the natural condition of our marine ancestors was not solitary but surrounded by other organisms, and that the evolution of the immune system arose as a way of recognizing and recruiting desirable living partners; the exclusion of pathogens and undesirables, which medics assume is the essence of the immune system, was essentially an evolutionary side effect.

The fungus *Candida albicans* is a normal inhabitant on human skin but feeds, as do *Staphylococcus* and the bacteria that create dental caries, on the massive amounts of processed white flour and sugar in the modern industrial diet. When such fungi and bacteria overgrow their normal borders, they cause *perleche* (clown mouth, or being "down in the mouth"), vaginal yeast infections (a yeast is simply a unicellular fungus), and other ailments that can be treated by strengthening the immune system and ingesting *acidophilus* and other friendlier symbiotic bacteria.

All visible life-forms are symbiotic—they contain multiple life-forms. Your cells and tissues have bacteria and viruses living in and on them; battalions of fungi, roundworms, and pinworms are normal inhabitants of healthy skin. The stomach and intestines contain immense crowds of yeasts and bacteria—"germs" that not only help digest food but provide trace nutrients and vitamins such as B_{12} and folic acid that "we" could not produce without "them." Recent experiments reveal that obese people lack certain types of helpful bacteria in their gastrointestinal tract. Any organism that can alter our psychology, even if only statistically, in a way that promotes its own propagation may promote itself. Far from being harmful, such organisms may augment our own survival, even if it means changing our behavior, cell composition, and genes along the way. Indeed, the most effective alliances, the deepest mergers, run so smoothly that they tend to go relatively undetected. Illness is not the presence of a foreign organism, but an alteration of the complex dynamic equilibrium among organisms whose normal community operations are synonymous with health. The organisms that spread by inducing or enhancing coughing, sneezing, sex, and so forth have happened upon ways to augment their own survival. The less we notice them, the greater the chances they are integrating, becoming us.

The problem of isolating causes in complex systems such as organisms is immense. For example, although we know that cigarettes are strongly correlated with mortality, the oldest woman on record, Jeanne Calment, smoked two cigarettes a day and lived 120-odd years. But she also drank wine, lived in the healthy countryside of southern France, rode her bike daily, ate a low-fat Mediterranean diet, and had a splendid sense of humor. The relative influences of various factors are difficult if not impossible to scientifically determine. Nonetheles, the symbiotic possibilities, like the sea in the Iggy Pop lyric, are endless. The relatively recent phenomenon especially prevalent among young women of systematically picking at or mutilating the skin, so-called cutting, can provisionally be traced to present suburban isolation in contrast with the earliest human habitats. Human evolutionists argue that our ancestors were hairier, like our closest living genetic relatives, the great apes—the gibbon, the chimpanzee, the gorilla, and the orangutan. The social behavior of the earliest humans, like that of modern chimpanzees, sometimes centered on the mutual removal via fine motor skills of sometimes fatal disease-carrying insect parasites. Worldwide, ticks and lice have carried typhus fever, and fleas the plague. Encephalitis, Lyme disease, leishmaniasis, sleeping sickness, and Chagas disease are among the modern insect-borne diseases, and the warmer tropical regions tend to have more. Presumably, grooming behavior—the plucking of insects from another hominid's hair—not only was pleasurable as a sort of proto-massage or back tickle, but also conferred substantial survival value by using manual manipulation to lower the incidence of disease in grooming populations. Comparative anatomy suggests that our ancestors were far hairier; the social behavior of grooming must have been strongly selected for in ancestral, tropics-living hominid populations. The loss of hair among human ancestors would have lowered the incidence of insect-borne diseases, but the evolved behavior for removing them, grooming, could have been "exapted": The fine motor skills evolved in the context of health among the hairy was perhaps a precondition for tool use, knitting and sewing clothes, building arrowheads and spears, among the unhairy. It is interesting to speculate that the origins of something as grand and planetary as technology may have had humble origins in cross-species hygiene. The ghost of other species may been instrumental in the origins of technology. Our ancestors' successful fidgeting, their primeval dexterity, was retrofitted for finer things after the development of eye–hand coordination, and the evolved manual skills and rare sense of touch were let loose in the nonliving world. Fingers and thumbs came to alight on things other than hair-clinging arthropods.

The consensus view is that the loss of hair in the naked ape was directly due to the terrific advantages that such loss offered runners and hunters. Human beings remain the fastest runners among all mammals over long distances; even the cheetah, so fast in the jungle dash, cannot keep up with man in the long run. This great hunting advantage, so pregnant with our species' future murderous history, was made possible by the hair loss that let our ancestors sweat during their long-distance hunts. But the powerful, earlier-evolved grooming behavior, a binder of social ties and a prelude to lovemaking—as well as to tool use—continued as a pleasurable social cement after hair was gone. Today, in our relative furlessness, we continue to crave the fine motor touch of the ancestral groomer and parasite remover. The pathology of the urban or suburban "cutter"—surrounded by people but untouched and alone—manifests as self-grooming where there is neither disease-causing insect parasite nor hair. In short, pathological plucking and cutting may have a deep evolutionary as well as a surface psychological explanation. The ancestral environment is gone but our evolution to it remains. Our lives are made in the image not only of present but also of ancestral others.

In the evocative thought experiment of microbiologist Clair E. Folsome, if your animal cells were to magically disappear:

> What would remain would be a ghostly image, the skin outlined by a shimmer of bacteria, fungi, round worms, pinworms and various other microbial inhabitants. The gut would appear as a densely packed tube of anaerobic and aerobic bacteria, yeasts, and other microorganisms. Could one look in more detail, viruses of hundreds of kinds would be apparent throughout all tissues. We are far from unique. Any animal or plant would prove to be a similar seething zoo of microbes.

We are surrounded by unsensed sensitivities and distributed intelligences. Actinobacteria in the woods make the forest smell like the forest; algae in the ocean, releasing dimethyl sulfide, make the sea smell like the sea. The earliest technology may have evolved in the crucible of ape–insect interactions, just one of many sorts of interliving upon a planet spilling over with bacterial, fungal, plant, and animal life. A planetary surface so full of life that it may itself usefully be regarded as alive.

Biospheric or Gaian, the resuscitated ancient notion of a living Earth is not a crackpot idea.

It is in the air.

Energy from the sun
that runs through all life is ultimately a Pandoran excess that cannot be closed up and kept tidy. The global environment, like Rome in its senescence, will always be open to organisms evolving new ways to plunder it. We must now restrain our growth at the cost of other species not because of any intrinsic evil but because those other species, in combination and with the environment, support us. Other organisms have preceded us in restraining themselves from growing so fast they imperil themselves. "Power," Nietzsche observed, "makes stupid." Our ability to tap into natural resources has promoted our global growth but nudged us in the direction of climate collapse. To avoid plunging, we must plunder nature not as a materially limited resource base, but as an inexhaustible bank of experience and information.

Evolutionary and ecological studies, which tell us how life may have survived former biospheric crises, as well as what is going on exactly with the surface of the Earth now, may be crucial to our future quality of life, if not existence.

To make it in the long term, it seems we should model our activities after the ancient survivors of this planet. And the thing that jumps out when we look at ecosystems is their interconnectedness and diversity. Although so close to our heart, it may be that cities—Gandhi's problematic Western civilization—are, from a global and evolutionary perspective, equivalent to disease. True, the biosphere is not like an ordinary organism that has a discrete form and a typical appearance. It is a constant mutant; its growth is open-ended. Just as, when our bodies are overrun with *E. coli* beyond the numbers normally found in the gut, we get sick to the stomach and can even die, so normally neutral human beings, reproducing rampantly, can cause disease-like symptoms in the biospheric body politic.

Nonetheless, we seem to be on the verge of creating an important, never-before-seen biospheric function: development of the technical means for Gaia to cart her offspring into space.

I wrote a book, called *Biospheres,* in which I argued that Gaia was not only an organism but also, as recent efforts by humans in building closed ecological systems seemed to indicate, on the verge of reproduction. My motivation was simple: I wanted to "prove" that Gaia was an organism to those who objected that she couldn't be because she didn't reproduce. As I researched, I found that the University of Hawaii ecosystem biologist Clair Folsome, quoted above, had kept symbiotic populations of multicolored bacteria growing in laboratory glassware for decades, since 1967. In principle the closed ecosystems he confined to their test-tube worlds might live forever. From Ecosphere Associates in Tucson, Arizona, I received a glass ball five inches in diameter containing seawater, bacteria, algae, and six pink shrimp—a world supported on a plastic tripod. I read Ian L. McHarg's description of the minimal self-sufficient living system. A Scottish-born environmental architect and landscape designer from the University of Pennsylvania, he pondered "how an astronaut might be sent to the moon with the least possible baggage to sustain him." The experimental environment dreamed up by McHarg consisted of a plywood capsule with a fluorescent light, air, water, algae growing in water, bacteria, and a man. With this most "modest hoard of groceries,"

> [the man] consumes oxygen and exhales carbon dioxide; the algae consume carbon dioxide and expel oxygen into the air which the man breathes, and so an oxygen–carbon dioxide cycle is ensured. The man thirsts, drinks some water, urinates, this passes into the water medium in which the algae and bacteria exist, the water is consumed by the algae, transpired, condensed, the man drinks the condensations and a closed cycle of water exists. When hungry, the man eats some algae, digests them, then defecates. Subsequently, the decomposers reduce the excrement into forms utilizable by the algae, which grow. The man eats more algae, and so a food chain has been created. The only import to the system is the light from the fluorescent tube. . . . The only export from the system is heat.

"Alas," concludes McHarg, "experiments of this kind have not been sustained for more than twenty-four hours, a sad commentary on our understanding of human nature."

One unsuccessful early Soviet model, called *Bios*, even contained a minimalist system similar to the one described by McHarg, but the chlorella—a hardy, protein-rich unicellular alga used to purify the water and air—gave off too much oxygen and proved to be less effective than plants at cleaning the air. To make functioning biospheres, the food cycle cannot be cut as short as it was in these early-1960s models. On November 11, 1983, two Siberian engineers, Nikolai Bugreyev and Sergei Alexeyev, entered *Bios 3*, the third Soviet version of a completely enclosed cosmonaut capsule. While provisioned with only one month's supply of food and separated from civilization except for electricity and television, the two researchers survived the Siberian winter, emerging from the enclosure's Earth-bound hull some five months after the metal door had been shut. Later models in the Bios program were nearly the size of *Skylab* and contained up to thirteen square meters of planted area per researcher, with dill, peas, wheat, kohlrabi (a cabbage derivative), and many other vegetables growing under xenon lamps, whose spectral light-emission properties are similar to those of the sun. Bypassing the soil, the vegetables grew hydroponically in specially prepared, nutrient-rich water. The stated purpose of the Soviet *Bios* projects was to reproduce on an interplanetary ship all the pertinent biology that occurs on Earth. A spacecraft or community in orbit can be continually resupplied by cargo ships, but soaring (or diving) away from Earth requires freedom from merely "camping out" in space.

The ideological pioneers who were planning on building the world's biggest artificial closed ecosystem wanted their story told. They had contacted ecologist Norman Myers, but he was too busy. I was thus the second choice to chronicle the formation of the notorious Biosphere 2 in the desert surrounded by the beautiful Santa Catalina Mountains outside Tucson. The visionary NASA astronaut Rusty Schweickart, also a guest of the Biosphere 2 crowd, said:

> The grand concept of birth from planet Earth into the cosmos—in 1993, 1994, 2010, 2050, or whenever—is a calling of the highest order. I want to pay a lot of respect to everyone associated with that grand vision for their courage to move ahead with this in the face of the unknowns which make the lunar landing look like a child's play toy. There were a lot of complexities there but we were dealing with resistors, transistors, and optical systems which were very well understood. Now we're wrestling with the real question: that natural process of reproduction of this grand organism called Gaia. And that's what all the practice has been about.

I agreed with him. This, "Biosphere 2," as they hubristically called it, was to be the biggest human-containing ecosystem of all. After going to the site before completion of the construction, and meeting with some of the developers and "biospherians" who were to be shut up inside, I still had my doubts. But I concluded that, whether or not Biosphere 2 succeeded in its lofty goal, such projects were not random. Rather, they indicated that Gaia was on the way. It was the fate of the Earth to reproduce. The biosphere, as unconsciously as a pregnant teen is unconscious of the details of her own complex reproductive biology, was on the verge of reproduction.

As I dug deeper, I found it significant that, however arrogant we were as a species, under current technology there was no way we could survive in deep space alone. We needed the ancient room-temperature enzyme-based recycling technologies of species in combination to survive. Other species, especially of bacteria and fungi, would be needed to recycle human waste back into food. Without them, there would be no prospect for long-term survival of the human species in space. This underlined to me the correctness of the Gaian view that planetary life is a single system, not merely a random collection of species living in different environments.

I found that research by Soviet biologist M. M. Kamshilov supported the idea that multispecies Gaian assemblages, rather than animals, were the "highest" evolutionary unit. Kamshilov had tested the resistance of a variety of communities to the toxic effects of phenolic acid. After experimenting with hydrogen and benzene, the Nazis settled on phenolic acid for giving lethal injections. Although harmless in small amounts, phenolic acid was used as a poison; at first it was injected into people's veins, later directly into their hearts, in Nazi concentration camps. Kamshilov (who may have researched the compound because of its history as a poison) found that the most complex assemblages most quickly broke down this potentially hazardous substance. Ironically, in smaller amounts, phenolic acid may be quite good for you: it has antioxidant and anticarcinogenic properties, and may even be a reason that fruit and vegetable intake is inversely correlated to heart disease and cancer.

Kamshilov added phenolic acid to four different model ecosystems. The first consisted of bacteria, the second of bacteria and aquatic plants, and the third of bacteria, aquatic plants, and mollusks. The fourth consisted of bacteria, aquatic plants, and fish. Although only bacteria can break down phenol into its harmless constituents, the latter-mentioned communities containing fish and mollusks as well as bacteria and plants disintegrated the toxin the fastest.

Kamshilov cites several factors for this phenomenon, including the fact that the waste products of the marine animals are food for the bacteria, thus speeding up disintegration. He also notes that as plants and animals were added, microorganisms that feed on bacteria were included. These microorganisms normally adhering to plant and animal bodies returned the mineral elements used in bacterial growth to the medium, further speeding detoxification. Kamshilov concluded that "the greater the diversity of species the more vigorous is the destruction of the toxic material."

These results seemed to indicate that the biosphere, that is, the global ecosystem, although it might not be exactly an organism, might be something *more* than an organism. Looking into the crystal ball that was my ecosphere, with its tiny shrimp feeding on algae, I saw that, just as throughout human history common areas became ruined through overuse, so the atmosphere and seven seas would be polluted—dictating the deployment of biospheric technologies on Earth long before they would be used for long, long-term space voyages or extraterrestrial colonies. If land, among the relatively crowded Europeans, had become private property, so air pollution would mandate production of closed ecological systems. Without them humans would suffocate in their entropic waste, much like the cyanobacteria before they evolved to tolerate their reactive oxygen waste. Our evolution, lagging our inventiveness, would take the shape of mass-produced biospheres, perhaps similar to Biosphere 2. From space our technologically developing planet would take on the appearance of a developing embryo in its early, spherical phase known as the blastula. Although more than an organism, the Gaian macroorganism would develop in the womb of space. The Biosphere 2 project was the clearest sign yet that Gaia was more than an idea. The biosphere was real, and it was beginning to bud.

As it turned out, Biosphere 2 failed. Peter Warshall, the former chief editor of *Whole Earth Review* and a consultant on the structure's simulated rain forest area (it rained from the ceiling), pointed out that one of Biosphere 2's problems was that its pollinating insects died out, leaving the plants that reproductively depended upon them unable to reproduce—and the humans who wanted to eat them gastronomically impoverished. The ecosystem experiment revealed that trees require wind to grow upright. Deprived of wind indoors, they lost their orientation, twisting and turning their branches too close to the ground. Moreover, the project wasn't exactly what was promised. The costly basement, importing electricity to keep the air system running in the would-be architectural icon of self-sustainability, resembled the engine room of the *Queen Mary.* The project had been funded by money from Ed Bass, described as the scion of a wealthy Texas oil family. The protective founders, apparently interested in possible patent applications and worried about a sustainable shelter in the event of a nuclear war, were inimical to the scientific ideal of shared inquiry and results. On the bright side, Biosphere 2 might help hone the prototype for recycling space capsules, and it could serve as a biospheric laboratory and example. Like the cute orange shrimp swimming in their little paperweight world, it demonstrated the intimate link between biodiversity and individual survival. Species do not go it alone. Ecosystems, unlike the *Apollo* astronauts on their short, three-day trips to the moon between 1968 and 1972, do not need to carry their food with them. They are their own food, and their ancient technology, low and ancient as it may be, promised to be the future high technology of spacecraft by becoming the gardens necessary to longer trips to Mars or beyond.

Bisophere 2 had expandable lungs connected to but set apart from the main edifice, which was broken into different biological areas or "biomes" each mimicking a distinct ecological setting: ocean, grassland, rain forest, marine, human–technological, and so forth. The lungs expanded when the sun came up in the Sonoran Desert, winged by the magnificent Santa Catalina Mountains. At dusk, as the sun suffused the sky and mountains in pink light as grandiose and maudlin as the hubris-ridden human structure it illuminated, the lungs in their separate but connected building would contract, allowing the main structure to deal with the natural pressure changes that might otherwise pop or deflate a sealed building in the desert.

As luck or planetary connectedness would have it, my own book, *Biospheres,* was kept in the library, which could only be reached by mounting a spiral staircase—not an attractive option for the original inhabitants, not only because of that book's bizarre thesis and overweening prose, but also because, in an architectural oversight, the oxygen in the atmosphere had reacted with the cement making up the base of the structure. The Biospherians, intrepid voyagers to ecosystemic inner space, were dangerously deprived of the very gas that had long ago poisoned Biosphere 1. The first group of Biospherians, which included longevity expert Roy Walford, were suffering firsthand from the inferiority of man-made relative to natural physiological intelligence. They were hungry, light-headed, and on the verge of asphyxiation. Reports had it that they were starting to resemble concentration-camp victims. Climbing the stairs to read my book would not have been preferable to sitting cross-legged in the jungle biome, watching water droplets collect before it "rained" from the ceiling.

Although it seem natural to us, since we use it to see out of our enclosed buildings, glass may not be the best material to contain an ecosystem. The term *greenhouse effect* describes how certain transparent gases—principally carbon dioxide, methane, and water—warm by letting in light but then trapping it as heat. This of course is what a greenhouse, an enclosure of transparent glass, does. But Earth's biosphere, after all, is contained not by glass but by gravity. And air.

I'd like to think

that the combination of space travel and the technology of building new ecosystems opens up the possibility that life can grow to inhabit the entire universe. The evolution of humankind creating in reality, through technology, the bliss it had glimpsed darkly through the veil of religious worship and wishful thinking. As one of the main theorists behind the effort to build human-containing artificial biospheres, John Allen, has said, "An extraordinary range of efforts by individuals and institutions is needed to make this transition into our solar system home, humanity's first great step to the stellar world." As British-born American physicist and mathematician Freeman Dyson has intimated, once life gets a toehold in space, it may kick off its human shoes. Though it hasn't happened yet, biospheres reproducing in space, dotting the universe with life, mixing with extraterrestrial minerals, synthesizing food, perhaps, from raw atoms, marks the beginning of an epochal process, namely life's cosmic evolution beyond the Earth. Once started, there is no telling what may happen. Life may ultimately merge its signifying codes and organic instructions with the electromagnetic spectrum, realizing that great science fiction fantasy, the transformation of living matter into divine light. Assuming that light isn't already alive; who can say where life ends and the universe begins in a world of organisms exchanging materials with their surroundings to survive?

Yet there is a troubling alternative. I've heard it referred to with vague accuracy as "the black thing." Talking outside in April 2004 in the beautiful Villa Serbollini conference center on the promontory of Lake Como, in the village of Bellagio, Italy, lighting one cigarette with the end of another, German biologist Wolfgang Krumbein shared with me his notion that the most deserted of all spaces—the relatively vast expanse between the nucleus and the electron—was the real-world cognate to the cosmological emptiness we sometimes viscerally feel: "the black thing." Even if the universe were to shine bright with the planetary brood of future biospheres, there is no assurance that the rapid expansion of stars will not forever prove a fatal impediment to life's natural designs of celestial expansion. There is sometimes something almost mocking about the tininess of humanity in comparison with the universe. It is as if ultimate reality were giving itself the slip, hiding its true nature from itself. But perhaps this is mere human paranoia, and the scientific reality is that on a level too vast for us to understand, we contain, or resonate with, or reflect, or epitomize the whole of existence. Is it possible that, as some who have examined both quantum mechanics and philosophy have speculated, the external realm studied by science and the inner world we experience are in some deep correspondence? Are both projections of another, more fundamental reality? Or does the experience and intelligence upon which we continuously silently congratulate ourselves in fact belong to the whole of nature, perhaps even to "inanimate" particles? The absurdity of personifying the inanimate universe may be no more illogical than maintaining the conceit that we alone among nature's entities have volition and perception. Each March the chiefs and the high priests of the Hopi people pay their respects to the wind god, Yaponcha. In folklore the people went suffocating and overheating after Yaponcha, in a prank, was trapped by cornmeal mush being stuffed in a crack in the black rock on Sunset Mountain where he lived. After the crack was unsealed Yaponcha was released to make the soft clouds move again without everything blowing away; the people rejoiced at the gentle coolness they had taken for granted. While, from a scientific view, the Native story seems overly to personify nature, from an ecological standpoint it displays an admirable culturally embedded understanding of the depths of our link to this invisible medium.

And this element, archaic, quaint, invisible, and mythical, surrounding you and perturbed to make the sound waves that carry your name, can truly be a secret power. In short, let me pay homage to it, aspirer extraordinaire, by telling you how to do a trick that requires no sleight of hand. In fact, it was the major miracle performed by James Hydrick, a man once featured on the TV show *That's Incredible* and billed as "the world's top psychic" by the tabloid newspaper *The Star*. His trick? Making a pencil, balanced precariously on the edge of a table, mysteriously turn seemingly by mental powers alone. No hidden threads or wires were used, and Hydrick appeared to use some weird martial arts technique to make the pencil move. He also could make telephone book pages mysteriously turn. The secret? Not sleight of hand but a gentle, human-blown wind.

Which is to say that Hydrick literally blew it. He used gusts of breath, directed not straight at the pencil but a little ahead of it on the tabletop, to propel it through an act of seeming psychokinesis. Mind over matter: Stephen King would be proud. Magician James Randi foiled Hydrick, who practiced in jail, by arranging Styrofoam pellets around the pencil so that when Hydrick blew, the pellets would expose his secret purse-lipped action.

On the TV show *That's My Line* with Bob Barker, Hydrick failed to do this simple trick, claiming his mental powers had temporarily departed. Too bad, because Randi had a ten-thousand-dollar check waiting for him if he could make the pages turn. Although a scientist and electrical engineer from Utah had already concluded that Hydrick's powers were real, Randi was prepared to offer the check because his crew had put a "shotgun" mike, which picks up sound from a very small-angle area, in place during the rehearsals, so they knew in the control room what Hydrick was doing, having heard Hydrick's propulsive bursts of air come in loud and clear. But Hydrick detected the mike and wouldn't allow its use.

As Randi explains,

> What of the pencil and page tricks? Well, my jaundiced eye recognized these as rather tired old tricks. . . . Hydrick was simply blowing the page over, and he spun the pencil around by the same means. Not immediately evident are these facts, however: First, the blast of air from a half-open mouth takes time to get to the props, and Hydrick made sure he turned his head away from the pencil and the page after giving a sharp puff of air, so that he was facing away when the action occurred. Second, one blows not directly at the prop but at the table surface.

When Randi did the pencil trick for me, it was impressive. You have to keep your mouth from moving, like a ventriloquist. (It also helps to have a beard.)

Not exactly savory, Hydrick was an interesting character. "My whole idea behind this in the first place," he said, "was to see how dumb America was."

> How dumb the world is . . . air currents . . . from my mouth. But you can't tell it because it took so many years of practicing to get this down pat to where you can't see it. I'm not just puffing out the air because that can be seen. I am taking the air from my inside and making it come out in a way in which it doesn't show. I can direct the air in a way that it hits head on every time. I spent one year and six months in solitary confinement. . . . I had spent hours and hours. I'd

> hold my breath. Different breathing controls. So many ways. I could make deputies think someone touched them on their neck because I could breathe in a certain way on their neck. They would feel something and say "That's a ghost!" They would piss on the floor and go running out of there! It was something that was fascinating to me and it got me recognition. I mean every deputy in that jail was so frightened of me. "That guy is possessed!" I remember when I was in the Chaplain's office. He taught me how to read and write. And I would convert people from bad to good. He told me that you had to turn them onto Jesus, the Lord. And he gave me a Bible and I'd read it. Then I got an idea! Now, I've never told Brother Joe this, and I've never told anyone this, but I would convert twenty inmates a day. That was my limit. I would have to convert twenty inmates a day. I'd get up there and start telling them about Jesus and stuff. And when I'd see that they were beginning to get turned off—I'd stop and say "You don't believe that it exists?"—I'd take a Bible and open it up and say, "If the Lord is here with me make these pages move!" or I'd open the Bible and say "Hold the Bible. Father in the name of Jesus Christ make these pages move." And the pages would move! And the guys are going "Oh my God!!!" Every time it worked. Then I would say "It's in you." Or I take a pencil and put it there and say I've got to call the Lord; but you are going to have the power to do this if you accept the Lord. The next thing you know you would see them with this big cross and handing Bibles out to people!

The use of air currents to make things move at a distance has other applications. An easy trick is to lay a cigarette on the table. You can pretend to polarize your finger with static electricity by vigorously rubbing it on your sleeve. Then leaning in, head down to conceal your mouth, move your finger mysteriously above a cigarette on the table—at a certain point you secretly blow, and the cigarette rolls.

If we move from considering ourselves the dominant species inhabiting a crowded rock, to a view of ourselves as populous components within a living system, the nature of coincidence becomes subtly altered. While it may seem so, it is not exactly blind good luck that sends a molecule of adrenaline from the adrenal gland into the blood, where it sets off a suite of physical changes in the response known as fear—although it may look like it if we do not see the wider, physiological context of the chemical interactions. So also seeming coincidences among human beings, and among human beings and the environment—coincidences that often seem strangely meaningful—may not be coincidental but part of the largely unseen complexity of a body-like environment.

The cycling order of a tornado or whirlpool, if seen without an understanding of the environment that energetically gives rise to it, might seem miraculous. The molecular arrangement is not random but coherent. The position of molecules, organized with a microscopic intricacy beyond human manufacture and fulfilling a natural function, might be mistaken as miraculous and purposeful. But this may be an illusion derived from viewing them separately and out of context when in reality they are part of a wave-like system, naturally organized to produce entropy through cycling. Within a complex thermodynamic system such as our biosphere, organisms and species can be expected, from one vantage point, to struggle for individual survival and to fend for themselves. From another vantage point, they are integrally connected to the bigger system to which they belong. Indeed, they cannot be understood without considering the larger system.

So it is that over time organisms and collectives that do not integrate their own needs with those of the living whole may drop by the wayside. For the moment we need not speculate on the mechanism, if any, that gets rid of organisms that don't promote the functioning of the whole. It is enough to notice that organisms today seem to have possible biospheric functions. Human beings, like other animals, ensure plant propagation and distribute bacteria and fungi to places they might not otherwise go. In addition, our global transportation and communication technology combined with the early forays into ecosystem science mentioned above suggests we may be evolving a function in propagating the biosphere itself.

It is difficult to understand how any species or, more generally, any kind of organism can become exceedingly widespread unless it conforms its activities to the entropy-producing whole of all organisms. Humanity, as a superabundant species producing toxins and meddling with the biosystem for its own ends, may be at a certain historical crossroads where our numbers either will or will not be tolerated. For our presence to be tolerated by the biosphere in current (six-plus-billion) or even greater numbers may require that we involve ourselves in the overall life of this planet to a greater extent. No matter how adept an organism is at propagating itself, if it does not fit the life cycles of others, it will fail. Entropy-producing forms such as ourselves with our massive collection and expenditures of energy may well be favored by the second law but not by life as a single system. The growth characteristics of organisms, sensing conditions and growing only within certain constraints, lead to internal feedbacks that have maintained biospheric life in the evolutionary long run. Although we may not understand global physiology in detail, the intricate feedback of the Earth system, combined with the fact that it recovered from each mass extinction to repopulate the planet with more species than had existed previously, suggest that it can react to compensate for perturbations that interfere with its smooth functioning as an energy system. Even human-induced climate change will not pose a threat to the system as a whole, although it may wipe us out. In short, we need to look at organisms, populations, species, and interspecies communities of organisms not only from the vantage point of their cellular and genetic parts, but also as organ-like systems within a functioning whole.

And the human requirement to become integrated into the biospheric system becomes critical as our presence becomes more pronounced within the planetary environment.

To explore how we might "become integrated in the biospheric system" and perpetuate our survival, consider the evolution of birds: Precisely their waste may have ensured them longevity. Coming from reptilian progenitors, birds could have precipitated an improvement in the cycling of an element vital to biospheric functioning: phosphorus, found in DNA. Birds may have biochemically "pushed" the biosphere, increasing global metabolism, the efficiency of the distribution of nutrients in a globally recycling system. It appears that birds not only did what it took for them to survive on their own, but were perpetuated because they found a niche within the planetary nexus, enhancing the connected activities of other organisms. They were "selected" not only by competition in various environments but also by Gaia, the global environment as a thermodynamic whole.

For a long time now there has been what amounts to a "phosphorus crisis" in the biosphere. Although it ranks twelfth in abundance among the elements in the Earth's crust, phosphorus appears mostly in the biologically unavailable form of minerals such as apatite, wavellite, and vivianite. Nonetheless, all organisms need phosphate ions present in fluids within living tissues. Neither DNA nor RNA can be made without phosphorus. The element usually makes up on the average only 0.000009 percent of seawater; this low concentration derives in part from the mad scramble of marine organisms to get the element when it is available as phosphate salts. The chronic shortage of phosphorus imposes an absolute limit on the abundance of life in shallow waters. Upwelling areas where deep ocean water comes to the surface often support rich concentrations of life. Such richness is due largely to the supply of phosphorus from dark waters where, in the form of a submarine rain of falling algal skeletons, it gradually accumulates. Taken out of circulation by the death of algae, the phosphorus floats with their tiny shells to the sea floor, limiting other organisms that would grow if only they had access to the element. Upwelling brings this vital resource back to the surface.

Phosphorus is needed to make not only DNA, but also other compounds crucial for the storage and transfer of energy and information in lining cells. Yet cells cannot assimilate phosphorus as apatite. Unlike other nutrient elements needed by life, stores of phosphorus on Earth are rare. Virtually the sole exceptions are the deposits found in some meteorites and in Cambrian and late Proterozoic phosphorite. Unlike bioelements such as oxygen and nitrogen, phosphorus is never found in a volatile or gaseous form that can waft through the global circulatory system of the gaseous atmosphere. Confronted with the biospheric immobility of phosphorus, organisms have long been in a position similar to that of Tantalus in Greek mythology: They are up to their chins in water with fruit suspended overhead, but whenever they bend the water dries up, and whenever they try to grab the fruit the wind blows the branch out of reach.

Birds may have alleviated this biospheric shortage by doing something often advocated in the human realm: They improved not the quantity of food but the means of distribution. Answering nature's call in the air, birds may have become biospheric messengers. Bird droppings—a major component of which is biologically assimilable phosphorus—long ago spread phosphorus on a global scale. Organisms were no longer tantalized by their inability to get at their food.

Phosphorus can also travel long distances as freight carried by swimming or crawling animals, or via microbial spores floating haphazardly in the winds. With the appearance of migrating birds, the equivalent to a biospheric proposal for phosphorus overnight express mail had been put on the table: Phosphorus was deployed, transported, and delivered; it was spread thin despite its refractory nature through the circulating atmosphere. Guano mounds on the islands off Peru may be thirty-five meters thick; sea birds and even bats transport it through the air, giving the slow-moving element "wings." Could seasonal bird migrations serve a relatively precise function of biospheric phosphorus distribution? The face of Janus looks both within, to the survival of organisms within populations of a species, and outward, to the role played in biospheric organization. Migratory fish such as salmon that swim upstream many hundreds of miles to breed may behave similarly in terms of helping distribute globally scarce but vital geochemicals. (And natural salmon populations may be under threat of extinction.) This is not to say that fish and birds evolved in order to transport phosphorus. Yet their serendipitous status as atmospheric couriers may render them integral to biospheric physiology in its present form—not as crucial as, but playing a similar role to, hemoglobin molecules delivering oxygen via blood in your body.

Termites may also play a role in the biosphere. The equivalent of one-third of the matter grown by plants each year is devoured by wood-eating termites and moves through their insides; much of this matter becomes the atmospheric gas methane. Methane is a more powerful greenhouse gas than carbon dioxide. It reacts with oxygen to make carbon dioxide and water. Does it—like trace chemicals in our own body—have a natural function?

Termites occur over about two-thirds of the Earth's land surface; there may be as many as three-quarters of a ton of termites for every person in the world. Moreover, termites convert roughly 1 percent of the carbon in wood into methane, in the process producing as much as half the methane in the atmosphere. Since methane may be necessary to stabilize the atmospheric level of oxygen, and since termites produce so much "natural gas," they may be playing a crucial role in the atmosphere, the biosphere's circulatory system. In short they may—like birds—be so successful because they help not only themselves but also the Earth system function as a living body.

Czech novelist Milan Kundera writes of human bias in history, wistfully suggesting that the migration of a given population of birds from one city to another ranks in importance along with European power struggles but is ignored because it is not human. Is this only the flip suggestion that history should be for the birds? Or is he saying that in focusing on the traditional human history of statues, wars, and kings, we have overlooked something far more important, the living history of the biosphere? Lovelock's second Gaia book, *Ages of Gaia,* describes the uncanny persistence with which the peacocks dwelling about his English country home gather together and deposit their smelly droppings on the pavement outside the door. At first, he said, he used to curse them, but then it occurred to him that those "ecologically minded birds were doing their best to turn the dead concrete of the path back to living soil again. What better way to digest away the concrete than by daily application of nutrients and bacteria in the shedding of their shit?" Just as the heart has its reasons of which the head knows not, so nature has its reasons of which humanity knows not. Many things we take as random probably have an ecological function within the interconnected system of life at Earth's surface.

The elegance with which disparate phenomena come together under the single explanatory umbrella of a living Earth is remarkable. It may be that the regulation of worldwide temperature, alkalinity, atmosphere composition, and the salinity of the oceans is only part of a more encompassing planetary physiology. Could there be, for example, such a thing as planetary psychology? I confess that I find the notion of a terrestrial intelligence appealing. For comparison, however unconscious, the mammal body—the underlying heart-beating, temperature-modulating, pheromone-detecting physiology of all of us—is in many ways far smarter and more dependable than the human mind. Scientific experiments have shown that whales communicate by songs that may be heard thousands of miles away beneath the ocean. However unconscious and physiological, a being as old as Gaia may have a very well-developed intelligence, even if only an intelligence of the body. And this bodily intelligence would exist on a scale of thousands of miles.

Keeping this view of a "geophysiological" rather than human intelligence in mind, we might wonder whether various forms of unusual communication (sometimes interpreted as "extrasensory perception") come about through biospheric channels, by unions too subtle to detect except insofar as they catch us by surprise. In principle, the atmosphere affords a biochemical conduit, with complex molecules detectable by many organisms. Could there be molecular messages sent or hanging in the air? Might the sort of meaningful coincidence that Carl Gustav Jung called "synchronicity" be due not to anything metaphysical but to the functional operations of the Earth as an extremely integrated organic entity? Once one attunes to them, many examples can be found, although it is difficult to prove that they are not mere coincidences. The most famous example, perhaps, is the case of the golden scarab that occurred with Jung himself, during analysis with a very rational female patient:

> Well, I was sitting opposite her one day, with my back to the window, listening to her flow of rhetoric. She had had an impressive dream the night before, in which someone had given her a golden scarab—a costly piece of jewelry. While

> she was still telling me this dream, I heard something behind me gently tapping on the window. I turned round and saw that it was a fairly large flying insect that was knocking against the window-pane from outside in the obvious effort to get into the dark room. This seemed to me very strange. I opened the window immediately and caught the insect in the air as it flew in. It was a scrabaeid beetle, or common rose-chafer *(Cetonia aurata)*, whose gold-green colour most nearly resembles that of a golden scarab. I handed the beetle to my patient with the words, "Here is your scarab."

To take a personal example, I had a dream that an unfamiliar relative approached me with the comment, "These Sagans seem to be everywhere, like stray jazz notes." I was embarrassed that I could not remember him though I thought I did detect a family resemblance. It turned out his name was Greg (I don't know any in real life), and behind him were more of the half-familiar brood, none of whom I'd met before. Not two days later, while awake, I turned to a Yahoo! Finance message board and found a quote from my father, Carl Edward Sagan, as the first, basically non sequitur entry on the NVEC stock board: "One of the saddest lessons of history is this: If we've been bamboozled long enough, we tend to reject any evidence of the bamboozle. The bamboozle has captured us. Once you give a charlatan power over you, you almost never get it back." The next respondent, unable to see the relevance of this post to a discussion about the prospects of a company making a new kind of magnetoresistive computer chip, wrote, "You talking about the Bush administration?" Somewhat surprised, I turned to another stock board, for Akamai, and the first post there also had my last name as the title. (The CEO is named Paul Sagan.) It was then that I recalled the dream. Now, I'm sure it could be pure coincidence, and it probably is, but perhaps it isn't. The world is not only stranger than we imagine, it's stranger than we can imagine. Of course, my father didn't say that: J. B. S. Haldane did.

Dwelling inside a living organization that dwarfs our own in size and complexity, what seem like suspiciously coincidental encounters may in fact be orchestrated, unknown to us, by the planetary macroorganism. As author Peter Russell points out, a sensitive cell in our body would no doubt often be startled by the peculiarly appropriate and repetitious order of its nurturing surroundings. Never having gleaned, due to its tiny relative size, its role within a functioning animal body, the hypothetical perceiving cell would be in an almost perpetual state of wonder.

Are we noetic inclusions, the equivalent of brain and sensory perception cells within the biosphere? Were the medieval purveyors of the microcosm–macrocosm analogy right after all, except that instead of the universe being a cosmic human, we are organelles within a thinking biosphere?

Historically, in the West, the stewardship of Earth has been considered a godly gift for the benefit of man, the chosen creature. Evolutionarily, provisioning for our life is explained under the rubric of adaptationism: Our ancestors were adapted to similar environments in the past. If the Earth is alive, however, we are surrounded not only by an environment rich in just those foodstuffs we need to survive, but also by a being or Being in our midst, of which, in the last analysis, we are living parts. Despite our intelligence, we need no more know what this colossal asexual but reproductive giant is up to than a brain cell—which may experience some sort of sensation on its own—knows of the giant human colossus that is, say, driving her Ford to work.

It may be that the time has come, from a biospheric vantage point, for us to glimpse a new role for ourselves: no longer isolationists, selfishly rearing technology for our own ends, but integrationists—connectors and vectors of disparate parts of the biosphere—no longer murderers but intermediaries and matchmakers among the millions of species participating in the life of the biosphere. In such an altered outlook, technology, language, and science would be seen to belong not to the ephemeral species of humans, but to the biosphere.

Sir James Jeans, writing in another time (circa 1930), compared the phenomenon of life to magnetism or radioactivity: It was a property of the universe and, as such, was no proof (once we disposed ourselves of our unscientific anthropomorphism) of a creator. The

mysterious universe in which we live was one into which we "stumbled, if not exactly by mistake, at least as the result of what may properly be described as an accident." Writing between world wars before discovery of the role of DNA, astrophysicist Jeans had good cause to question the postulate of a creator: Had the universe been designed mainly to produce life like our own, one would think there would be a mite more of it. Made of common stuff, Jeans attributed the unique traits of life to the inclusion of carbon atoms able to make complex chemicals.

> An omnipotent creator, subject to no limitations whatever, would not have been restricted to the laws which prevail in the present universe; he might have elected to build the universe to conform to any one of innumerable other sets of laws. If some other set of laws had been chosen, other special atoms might have had other special properties associated with them. We cannot say what, but it seems *à priori* unlikely that either radio-activity or magnetism *or life* would have figured amongst them. Chemistry suggests that, like magnetism and radio-activity, life may merely be an accidental consequence of the special set of laws by which the present universe is governed. . . . Again the word "accidental" may be challenged.

If we dismiss every trace of anthropomorphism from our minds, he continues, there is no reason to assume that the present configuration of laws was not expressly created to lead to magnetism or radioactivity—both of which are far more prevalent in the universe as we know it. Life is too insignificant to be the centerpiece of creation. Jeans likens the prevalent belief in a creator God to the conceit of a mathematically proficient but crazed sailor who believes that, since the knots used for tying ropes in sailboats cannot be tied in a space of one, two, four, or five dimensions, three-dimensional space exists so that there can be sailors. Both arguments are stretches, Jeans points out, because the general conditions provide for much else besides tiny humans on the small inner planet of a medium-size sun

as well as knot-tying sailors. Life, he convincingly argues, has the character of an accident.

And yet the evidence of our senses as well as our human proclivities leads us to find meaning beyond that of sheer accident. We do not judge ourselves by our ridiculous size within the vast universe. And we sometimes have profound experiences, such as the coincidences previously mentioned, that suggest our environment and interactions are more than a matter of mere chance. I have more than once experienced a forceful interaction with a kind of alien but familiar being I took to be the Earth. Lovelock colleague Stephan Harding smilingly uses the word *Gaia'd;* that is, he uses *Gaia* as a verb to describe the experience of a direct human encounter with the overpowering biosphere. Rightly, he uses it in the passive construction so it is clear who is the more powerful, the hominid or the biosphere. One time, at a family wedding at my father's house in Ithaca, New York, I was Gaia'd by a confluence of factors, including being offered stuffed mushrooms by a French chef dressed the part as I sipped red wine in a mist on green grass over one of Ithaca's famous gorges. The wedding party's Olympian placement contrasted with its suburban comportment; we were literally among the clouds, one of which began to obscure the sun above and darken the waterfall below. It began to rain. I felt for the moment as if I were in medieval France—and just then I was told there was a call for me, which I ran in to take in the kitchen. There were no cell phones at this time, but my father's then-elaborate phone system with pulsing red lights and multiple lines threw me into a place of temporal perplexity, caught between the call of the pastoral past and the future whipping around the corner as if to offer me a preview of the shape of things to come. Combined with the fact that the call was in reference to an article on the biosphere, and I truly felt Gaia'd. I remember another time, entertaining guests and trying to talk on the phone, but standing dumbfounded as I was assaulted by an atemporal biogeological eidolon, the intensely stark image, outside of time, of our fateful cycling spherical planet in my mind's eye.

Such experiences
may seem more meaningful than they are. Or they may be the result of certain altered states that allow us to—using Robert Heinlein's word from *Stranger in a Strange Land*—"grok" this planetary entity. They may be, though there is little room in our culture to conceive them as such, direct apperceptions of the biosphere, genuine communiqués with the Gaian macroorganism.

Perhaps Earth is for the birds. I met Brown University professor of Slavic studies Alexander Levitsky in the beer-bottle-strewn basement of planetary scientist Jim Head. When I found out that he taught a course in science fiction, I asked Levitsky if he knew the story about how Philip K. Dick thought writer Stanislaw Lem was not a person but a front concoted by the KGB. He certainly did. A nativized Czech, Levitsky is a large man and a born raconteur. He met his beautiful Swedish wife on a blind date where he was described to her as an unmistakable Russian bear. Levitsky, his small hands folded gracefully across his belly, is a self-confessed aficionado of small birds, an appreciator of Russian linden berries, which he says are much better than American cranberries, as well as fine cheeses—whose diversity overwhelmed him upon first vistiting France, such was the contrast with Czechoslovakia, which under communism had only one kind of cheese availiable. Well, I mention him because after we had ascended into a small room to dine, he told a tale of ironic ecology. In the context of reminiscences about the bizarrerie of modern-day American politics, which has strangely morphed into the sort of bland mind control that was all the rage at the height of communism—after telling stories of gigantic billboards in Soviet Russia that said EAT CABBAGE—Levitsky regaled us with a story featuring small birds. It seems that Mao Tse-tung, the Chinese leader, had become alarmed at the habits of sparrows of eating seeds used for grain in agriculture. The little birds' feeding habits incensed him and, in retaliation, he used the full government powers of the Communist Party to support a successful extermination campaign against the offending sparrows. To this day birds are known to be strangely rare in China. Great hordes of peasants went into the fields in order to appease the leader of the people in his apparent wisdom. Although

they did not have shotguns they made such a racket that the sparrows, frightened to the skies, continued to fly overhead until they literally dropped dead to the fields from sheer exhaustion.

The ecological irony is that sparrows are apparently the only known predator of locusts—the mass-feeding form of grasshoppers. Thus, shortly after the plentiful sparrows' disappearance from the air, great swarms of locusts descended upon the People's fields. The fields were decimated, devoured far more completely than the sparrows themselves ever could have mustered. I love this true story with its comeuppance against human beings who with a sense of entitlement and steeped in ignorance would dictate to nature as if her cycles were as simplistic as the linear mind of man. Apart from the story's status as a natural Aesopian fable with a possibly Confucian moral, it serves as an advance warning of the growing danger of replacing stable diverse ecosystems with foodstuffs and monocultures meant for humans only.

In my former hometown there is a small airport. I like to take a turn off the main road that leads to the highway and get off there instead. It is like going back in time: the twisting trees, the dried mud and perma-puddles, the fractal branches clutching at the air like a memory. This time when I went, James Brown had just died. I remembered another time when Frank Sinatra had just died, and I was listening to his songs on the radio. I'm listening to French music now, smoldering leaves and lipstick traces, 1950s jazz. The way the sun hits the brittle rooftops of the man-shaped corn, gently dancing in the wind. The future is after me. I have been. I feel alive when I come here. So does nature. There is a freshness to the mist, a pungent aroma to the air. In places around here you can look at a scene and find nothing to tell you that you are not living a thousand years ago. As I make my automotive rounds along the bumpy dirt edge of the lone runway, the sight of a metal plane, as if only imagined (perhaps I heard a hum), drops into view. Then—perhaps it was scared of me—it doesn't touch down but straightens, not landing but tipping its nose and taking to the skies again. I could write a massive tome to civilization this way, on the sunset-bleached pink face of consecutive Post-its.

Like the insects, and bats, and birds that paleontologists now classify as dinosaurs—we and our airplanes and exploratory spacecraft may be the only real angels. Besides the planets and stars, of course. The fire that drives us comes from food, cellularly burned in the presence of oxygen to give us our energy. The jet fuel of planes requires oxygen gas freed from its molecular bonds by life using the energy of the sun—the most abstract thing in the universe, because, as Georges Bataille has said, although it is always there, you can't look at it too long without going blind.

FIRE

Flames will destroy everything
At the end of the universe
It may already be destroyed
—Pi Yen Lu
(Blue Cliff Record), tenth century

Burning bright

in the universe we call night, all life knows its passion, from mitochondria's mighty mission to lovers lit by the moon's reflective white. We warm our toes by it and run from it when it rages out of control, its licking flames removing oxygen from the atmosphere, thereby depleting its rate of incidence. It generates not just heat but white-hot light itself—a mystery of mysteries, one of the secret names of God, that which nothing flies faster than, the absolute basis of relativity theory, the clarity of which the Yaqi shaman distrusts as illusory, and near the speed of which clocks grind to a halt relative to observers. Light made of a thing (photons) that, like the spirit is supposed to be, is bodiless, devoid of mass. Such massless photons, relative to us, may be eternal. Energy is neither created nor destroyed. Such that, if we make the equation between God and light—"I am the Light," said Jesus—we can find backing in Einstein's theories. Matter can turn to energy and light, and thus, almost angelically, flesh to spirit construed as a thing with no mass. But no thing may be nothing, and the devil was an angel—Lucifer, which comes from the Greek for "light bearer." Creator and destroyer, initiator of friction, raiser of temperatures, principle of passion and agent of ire. Transformer and igniter, crackler and consumer, secondary producer of charcoal in the Earth's crust—which is first made, like love, by life. We see it aglow under the rocket ship roaring into space; in the fall of meteorites disappearing as they slip into our atmosphere of a summer's night, reminding us of our own bright, limited tenure.

Prometheus stole it from the gods. Heraclitus presented it as the ultimate principle, the unchanging arbiter of all change, the radiant core of becoming that does not itself change. Warmer and transformer, increasing the velocity of atoms in substances exposed to it. Christian eschatologists—those who study religious beliefs concerning the future—say it is the primary torture instrument in Hell, that mass murderers who repent and welcome the Lord into their heart on their deathbed may avoid it, while philanthropists who don't will be eternally toasted by its hyperpainful flames.

Reality may be less melodramatic.

Life on Earth seems to be a manifestation of the dissipation of the sun. As the nuclear energy locked in the atoms of our local fireball becomes light and heat, the irradiated Gaian system becomes more complex, helping to spread the sun's energy into space. The first law of thermodynamics says energy is never lost. It cannot be created or destroyed. The second law states that energy will always be losing its usability, going to more debased forms, making it more difficult to derive work from it. And heat is the end point of thermodynamic processes. The expansion of the universe may give radiation a place to go, allowing for the development of complex forms. The whole story of the universe, from intensely hot beginning in the microscopic point of the Big Bang, to spatial expansion and relatively cooler temperatures, looks like a long drawn-out explosion, a slow-dying fire, energy's great spread. In other words, both the universe and life appear to exemplify the second law of thermodynamics, which can be restated for open systems as energy's tendency to spread. This is a simpler and, as we shall see, more general and accurate description of the second law than the more familiar one, that the second law defines an inevitable increase in entropy—taken to be disorder.

Burn, baby, burn: Energy scientists in the nineteenth century saw the cosmos eventually fizzling out, dying down, and expiring like a waning fire in what they called the heat death of the universe. For the growing number of the educated, who looked to the experimental results of science as more loyal than the authoritarian stories of traditional religion—which told you what you wanted to hear, or terri-

fied you into believing its claims—there would be no respite from this future anti-Hell of lukewarm embers. It was after all built upon the experimental conclusion that complex systems became less so over time; that energy was always lost, becoming unavailable to do work. It was concluded, to paraphrase T. S. Eliot, that the cosmos would end not with a bang but a whimper.

The directionality implied by classical thermodynamics was difficult to deal with, seeming to expand the specter of cosmic death with the indomitable authority of modern mechanical science. It replaced both the optimistic story of Christianity with its possible happy ending in Heaven, and the classical viewpoint of celestial mechanics as a timeless friction-free realm that had gone on perhaps forever and would continue till kingdom come. Either way you looked at it, it was depressing. As the brilliant George Carlin has said,

> The most unfair thing about life is the way it ends. I mean, life is tough. It takes up a lot of your time. What do you get at the end of it? A death. What's that, a bonus? I think the life cycle is all backwards. You should die first, get it out of the way. Then you live in an old age home. You get kicked out when you're too young, you get a gold watch, you go to work. You work forty years until you're young enough to enjoy your retirement. You do drugs, alcohol, you party, you get ready for high school. You go to grade school, you become a kid, you play, you have no responsibilities, you become a little baby, you go back into the womb, you spend your last nine months floating. . . . Then you finish off as an orgasm.

This literal reversion of science's version of the Fall redresses cosmic or divine oversight, but it is a comic fantasy, not the tragic reality. In passing it casts a more-than-suspicious light on the not too awesome craft of a traditional God taken to be both beneficent and omnipotent. If he were all-powerful and good, he might well have come up with something at least as titillating as George Carlin. Instead, it looks like he got time backward. The direction seems designed to displease, adding weight to the heretical worldview of the Cathars, also known as the Albigensians, victims of a bloody forty-year crusade unleashed by Pope Innocent III. The Cathars, who flourished in the twelfth century in the Ariège region of southwest France near the Spanish border, preaching nonviolence, eating fish, and avoiding animal fat, argued that the body and the physical world (including much of the Bible) were manifestations of the devil. A science-fiction version of these events—only for its believer it was not fiction—occurred in the world of Philip K. Dick, who, triggered by the fish (a symbol for Christ) necklace of a female visitor selling magazines door-to-door, experienced what seemed to be a transpersonal memory of his true place in space and time. After this 1974 experience, Dick revised his view of time to believe that after the fall of Rome time began going backward. In his last three science-fiction books, as well as in a largely unpublished manuscript of more than a million words referred to as the *Exegesis,* Dick tries to make sense of his divine visitation, which he reported partly took the form of beams of pink light and described as an encounter with a transcendentally rational being.

Mystics and heretics aside, the direction toward dissolution described by the founders of classical thermodynamics mirrored the Christian description of the Fall without any scientific equivalent of salvation. It was a bleak perspective, but one based on experiment and evidence rather than a doctrine of primordial guilt.

Then along came evolution. As presented by Darwin, the historical change of forms on Earth's surface was a progressive process, flying in the face of both the thermodynamic view of inevitable dissolution and the religious teaching of original sin. The fossil evidence, rather, showed a development, over hundreds of millions of years, of ever-more-advanced and impressive forms, arguably culminating in Englishmen like Darwin. The universal story looked more like a rise than a fall. Evolution, although it acknowledged our embarassing, apish roots, gave a new sort of centrality to the human being, who could now be viewed as the most evolved being upon a planet clearly destined for higher things. Like the thermodynamic view, whose thunder it stole, Darwinian evolution presented a scientific alternative to eschatology, the religious direction of story defined by the Bible. Evolution, like the Bible, was temporally linear, telling a story with a reasonable beginning and probable end. But evolution, unlike thermodynamics, seemed not only off to a reasonable start but headed toward a happy end, as well. Now even the scientifically minded and unsaved could garner hopes for better times ahead. In contrast to thermodynamics' view of a heat death, evolutionists could exult that, in the words of a banner I saw at a science-fiction convention, THE FUTURE AIN'T WHAT IT USED TO BE.

But the notion

that the second law of thermodynamics guarantees a one-way trip toward disorder, disorganization, and death, and that it thus contradicts the evolution of life, is plain wrong. Despite their different eschatologies, their seemingly different stories and directions, evolution and thermodynamics are not opposed. Both are based on scientific evidence and so must, in principle, be reconcilable. In fact, as we have already seen, they are: Darwinian systems are thermodynamically driven, their cyclical complexity and growth connected to the navigation of energy streams and production of entropy as heat.

Nonetheless, the idea that evolution violates the second law is still popular, in large part because it supports creationist arguments. If the second law mandates disorder, and evolution displays order, creationists claim, then life cannot be natural but must be supernatural. The creationists want to claim that their view is science but, of course, science by definition deals in natural, not supernatural, explanations. A view that life is not natural will forever remain ostracized from the intellectual honesty of open scientific inquiry. In fact, saying something is supernatural is like saying perception is extrasensory: Perception, even if we don't know how, occurs through some sort of sensing, just as the unexplained, although seemingly miraculous, belongs by definition to nature and the universe defined scientifically as all that is.

Nonetheless, I am fascinated by a concept I call the Todorovian—after literary critic Tzvetan Todorov, who pointed out that some phenomena seem to simultaneously admit of both a natural and a supernatural explanation. I am among those who believe that science enjoys a special privilege among the ways we have of understanding the world. However, science does not exist in a vacuum. Science is a human endeavor. And humans are fallible, even the pope. Moreover, modern science presents the world as the result of brute interactions among unfeeling particles. Yet we feel. We feel as if things matter and as if they are heading somewhere. But to what purpose, to what end? How do unfeeling particles make feeling beings? Perhaps the particles themselves feel, or the cells made of them, of which we ourselves are made. It is possible to imagine a world of actions and reactions without observers, without consciousness. It would be just like this world—the world described by science—except devoid of feeling. So it seems that, unless it is a useless illusion, feeling not only exists but also it means something. We could act as we do, reproduce and "live" without knowing that we do so. So perhaps we feel and experience for a reason, although we don't know what it is. Perhaps it is because the universe wants to remind itself of its true nature, and it must be divided into parts that perceive the other parts in order to do so. That makes sense to me, although it is not exactly scientific.

The particular stories of organized religion may be too unlikely, too far-fetched—too much like bad science fiction—but science alone is too mechanical. In philosophy this idea that we could all be here doing exactly what we're doing without knowing what we're doing or feeling ourselves doing it is sometimes known as the zombie problem.

The zombie problem in essence calls our attention to the superfluous nature of awareness. The gap between conscious and unconscious beings is so great, it suggests that consciousness can no more appear from a random interaction of particles than can a rabbit "really" appear from the dark shadow at the bottom of a top hat. As philosopher Charles Birch explains, in the context of the scientifically orthodox but ultimately challenged view of mind "emerging" from matter:

> At one time gaps in the fossil record were presented as a case against Darwinian evolution. For most palaeontologists this

> is no longer a major problem (see Maynard Smith & Szathmary 1995). Questions about complex organs and questions about fossils are examples of questions that biologists ask and to which they can provide credible answers. They are the easy questions.
>
> But there are additional questions that are difficult and that Neo-Darwinism has up to now avoided. They have to do with intangible but essential aspects of the living organism. They revolve around consciousness. There is no problem in attributing survival value to consciousness. But why are we not zombies? Zombies are fictitious creatures devoid of any conscious experience and yet having behaviour identical to that of their conscious counterparts. They could get along all right provided they had inbuilt programs to avoid dangers and be attracted to favourable environments such as those that provide food and energy. I know of no argument in the mechanistic model of Neo-Darwinism as to why we and all living creatures are not zombies. What in principle is the merit of conscious behaviour *over* unconscious robotic behaviour of the same survival value? I know of none.

If zombies are like humans but don't really know or feel—if they sleepwalk, as it were, like unalive and unconscious machines—why is it that we have awareness in the first place? Are we not, in the end, chemical computers, a more advanced sort of natural robot or organic machine? Even emotions, such as fear, that enemy of happiness, can be understood as an algorithm for avoiding possibly deadly risk.

Many human activities, though they seem to derive from the depths of the soul (or, if you prefer, the primordial freedom of the quantum flux), seem instead to be simple algorithms. We are run by rules. The dog turns its nose up at what's in its bowl, a learned algorithm that continues because so doing in the past has garnered him gourmet rewards. The man seeks the curvaceous, healthy young woman, with skin unmarred by indications of disease—the physical inspiration for poetry and song long before the advent of knightly devotion, the romantic troubadours of Provençal, or Jonathan

Richman and the Modern Lovers. These mathematical-like verities ruling, or at least influencing, animal behavior are rote enough, simple enough, computer-like enough to imagine them being installed in an advanced robot, such as the advanced robots with whom P. K. Dick peoples his *Do Androids Dream of Electric Sheep?*

The robot can do everything we can, and less. At least in theory. The android or cyborg's mechanical nature, even if such beings don't yet fully exist, alerts us to the philosophical problem of our own awareness, which seems superfluous in a world of mechanized computer instructions, algorithms. What is different about humans is, perhaps, the extent to which we can rewrite our own algorithms. We can devise meta-insructions, cerebro-cortical guidelines influencing our habitual and even instinctive behavior. Brain scientists and, to a lesser extent, evolutionary psychologists call it neural plasticity. Its more familiar name is learning. Animals can learn, and change their behavior, but not with such a wild behavioral palette as we. Philip Dick likes to have characters who don't know they're androids as well as nonandroids—regular garden-variety hominids like you and me—that act coldly, algorithmically, and *could be* androids. Perhaps the best example of this in his work is the short story "Impostor."

The hero of Dick's story is Spence Olham. Olham learns that an alien humanoid robot, a cybernetic robotic weapon of mass destruction called a U-Bomb, has assumed his identity. It has been built to penetrate Earth's protective shield. Alerted and terrified, Olham searches the woods around his house and, lo and behold, finds a ship. It has crashed. Upon inspection it is seen to be damaged and to contain a mangled body. As Olham peers more closely he confirms the body's robotic status, which is revealed by a glint of metal in the thing's chest. Olham withdraws the metal evidence of his impostor. But as it turns out, it is not a piece of robotic innards at all: It is a "needle-knife"—the weapon that murdered the real Olham. *He* is the U-Bomb, implanted with false memories of being human.

The robot's realization makes him explode. Which explosion is visible 4.3 light-years away in Alpha Centauri, because it obliterates the entire inner solar system including, of course, the Earth. Mission accomplished.

Metaphysically, each of us may be a "U-Bomb" ready to explode with the realization that we are not what we seem. Matthew 5:48 quotes Jesus as saying, "Be ye therefore perfect, as your father in Heaven is perfect." But the word *perfect* in Aramaic means "inclusive." This may suggest that Jesus was advocating for us to be absolutely inclusive, to be one with the universe, which we tend to think of as lifeless and separate from our conscious, ensouled self. This may be an illusion, however, as may be the appearance of individual awareness from unaware particles. An alternative—entertained by Birch and other followers of the process philosophy of Alfred North Whitehead—is that experience occurs to some degree in all parts of nature. Awareness, in other words, is already there. It may be teased out, but it does not emerge magically from its absence. I must say as a magician that this makes more sense than the alternative. As mentioned, things may seem to magically appear on stage, but they don't. They are put there during the course of some distraction, or are already there and revealed by the removal of a secret cover.

For Dick,

empathy is the mark of the human. It is not the presence of cells, or the ability to wax profound at cocktail parties, or even self-awareness. Robots, as his stories (and modern computer scientists) suggest, may develop all of these things. What is human is the ability to relate to other beings, to imagine how they feel, to feel as if you are they. Ironically this hallucinatory ability—psychological identification accompanied by emotions whose very nature is to be shared, transpersonal—is a vector taking us out of ourselves. It is a route to demolition of the ego—the seeming source of who "we" are. So for Dick, empathy separates the human from the inhuman. But this same empathy takes us out of ourselves, making it possible to identify with other people, other races, the other gender, other animals, even trees and mountains. My ability to feel like you—to know what it's like to be a tiger, say—is in Tibetan Buddhism called *lojong*. It's a mystical art, shamanic. To step into another's shoes, to "other." The noted improvement in immune response, health, and happiness of the socially connected suggests an evolutionary function for empathy and interrelationship in populations whose members are generally more capable together than alone. As many have noticed and fewer practice, there is a unique paradoxical pleasure to the giving of oneself, to the helping of others. Empathy is a doorway leading out of the maze or prison of self. But since it is the path to the abnegation of self, that is, to nonself, it is a strange thing with which to define the human. If empathy is the essence of the human, then what is most us is that which allows us to be other. "I is another," sings Johnny Lydon of the Sex Pistols. "All organisms think they're human," writes Alan Watts. It is as if, when we step into the shoes of the shaman, we come to the startling realization that "the human" is itself—at least from the mystic perspective of vast, if not infinite time—an illusion. None of us is who she thinks us. You are, as Dick suggests, a U-Bomb. When the bomb explodes, you realize that you are not you but that. As the bumper sticker says, you are not a human being having a mystical experience, you are a mystical being having a human experience.

The world described by science could just as well be populated by chemically interacting, physical, law-obeying zombies. But it is not. We know we are here and feel as if our choices—beyond the determinism of mechanical science—affect outcomes. Theoretically, even minor choices in what we do today, in what we say or how we choose to perceive, may ripple through time, having an effect on the far-future universe. Again, assuming that our apparent freedom to choose is not a useless illusion. And even if we are not free to choose, what would be the purpose of such an illusion? Oddly, it seems we either have free choice, or are under the influence of someone or something who "wants" us to feel. I put *wants* in quotation marks because a truly supreme being may exist outside of time. Since wants and desires are oriented in time, such a being may not have desire, wants—or free choice—in the familiar sense at all. For him, her, or it—there is no proper pronoun, and the word *God* has been so abused that one can understand how the ancient Hebrews were reluctant to speak its name—everything already is. If everything already is, there is no choosing, which takes place in time. Time itself may be an artifact of the revelation of all that is—what Joseph Campbell called, in the context of Christianity as an example of the universal myth, the fall into incarnation and time.

The Todorovian is named, as noted, for the French-Bulgarian literary critic Tzvetan Todorov. He recognized a powerful category in literature he called *le fantastique.* The fantastic applies to phenomena that simultaneously admit of a scientific and a supernatural explanation. In a fairy tale or tale of sorcery we just accept the existence of the supernatural, at least insofar as the story goes. In other stories, what seems inexplicable turns out to have a supernatural explanation. For example, in Guy de Maupassant's short story "Fear," a man, traveling on a train at night, sees a creepy scene of men illuminated by fire in the woods despite the warm weather. This elicits a tale of another fearful night, in which de Maupassant's narrator describes his horror upon being lost in the fog and—in this time before automobiles—distinctly hearing the rattling of the wheels of a cart with no horse. Locating the source of the sound after some difficulty, the narrator, to his horror, witnesses a wheelbarrow roll on through the fog by itself. Only later does he realize that it must have been propelled by a child too small to be seen from his vantage point. This second story is an example of a seemingly magical tale that turns out to have a thoroughly rational explanation.

The Todorovian—I prefer this adjective, as it is unlikely to be confused with *fantasy*—is different. Some of Philip K. Dick's science fiction may qualify as *fantastique.* In one Dick story a broken-down spaceship is visited by a strikingly life-like Jesus image. The Jesus figure, however, proceeds to act like a vampire, biting into the flesh of the surviving crew. What at first seems to be a fantasy turns out to have an explanation. The hapless victims' cultural beliefs were accessed electromagnetically via their onboard computer by a local intelligent civilization attempting to offer aid. Unfortunately, the aliens, projecting through advanced technology the images of what they correctly divined was taken to be a deity on the planet of the visiting spacecraft, made the incorrect assumption that God would be an eater rather than eaten. The result, based in part on the irrational belief of the visiting hominids—what self-respecting God is eaten by inferior beings?—was an interplanetary diplomatic fracas. The point in the Dick story at which the barely surviving cosmonauts are visited by the alien presence of Jesus Christ—we don't know if it is a true visitation deep in space, or something else again—is aptly covered by Todorov's intriguing category of the *fantastique*.

I believe not just Dick's story but the whole world must be placed in this category. We don't know if the world is real or an illusion, and if it is an illusion whether it is a shared one. We need science but we must be open to what has not yet but may someday find a scientific explanation.

I don't have a problem with this. It means the world is fantastic. There is more in Heaven and Earth than dreamt of in our philosophy, which of course is only a tiny fraction of Heaven and Earth.

Saying the world is Todorovian, in the sense of the *fantastique,* is both a humble admission of our ignorance and a payment of respect to the scientific method. It is possible that a God of the sort described in the Old Testament exists—a sort of overgrown three-year-old with a beard and superpowers, given to creating scenes and exacting vengeance, with a taste for caprice. It is possible, but is it likely? Perhaps no more so than the Flying Spaghetti Monster, an alternative being proposed (in jest) to have made the Earth and humans by intelligent design. (For more on the Church of the Flying Spaghetti Monster, go to www.venganza.org.) In fact, despite the airtime given by the too often scientifically ignorant and politically meek media to creationists, the men who started the United States were deists, not fundamentalists: Their god, insofar as they had one, was on the side of science and nature, and not a supernatural nutcase operating beyond nature with jealousy jags, temper tantrums, and afterlife ultimatums. The argument for atheism is the same as the argument for monotheism except that there's one fewer god. This brings to mind Alexander Hamilton's reputed response to Ben Franklin's suggestion that each session of the Constitutional Convention be opened with a prayer: "We have no need of foreign aid."

While Americans may scoff at the beliefs of suicidal terrorists attempting to reap their reward of seventy-two dark-haired menstruation-, urination-, and defecation-free *houris* in paradise, the biblical

God is much the same—a deity who commands that you love him and threatens you with eternal torture in Hell (no matter what good deeds you may have done) if you don't fall for the transgenerational ideological scare tactic—that is, if you don't believe in him. But peer pressure doesn't make something true. Science is not about believing in something despite the lack of evidence—faith—but finding out the truth whether we like it or not. It is hard to believe in a God that, as my father says, has a gray beard, counts the hair on everybody's head, and tallies the fall of every sparrow. Such a God is simply way too human, testifying to the failure of the imagination of a being who, quite naturally, tends to see things in terms of his own, limited self. The Bible says that we are made in God's image, not that God is made in our image. Voltaire said that if God did not exist, it would be necessary to invent him.

A more reasonable intimation of divinity was developed by lens maker Benedictus de Spinoza, a Dutch philosopher whose family was kicked out of Portugal during the Spanish Inquisition because of their Jewish heritage. Spinoza's God was an entity partially contiguous with nature, a deity that did not interfere in time and therefore did not perform miracles or tamper with creation via divine intervention. His was a God as impersonal as the laws of nature were eternal, a God that had lost all traces of the anger management problems of the Great Steroidally Poisoned Despot in the Sky. Spinoza's God was the sort of deity to which Einstein referred when he used the term. Here was a God not separate from, but part of, the universe; a God overlapping with but extending beyond the universe, which had infinite qualities that were infinite in extent. It was not a personal God concerned in the business of punishing sinners or answering prayers. As Einstein wrote, "The most beautiful thing we can experience, the source of all true art and all science is the mysterious. He who knows it not and can no longer wonder, no longer feel amazement, is as good as dead, a snuffed-out candle. It was the experience of mystery—even if mixed with fear—that engendered religion. A knowledge of the existence of something we cannot penetrate, of the manifestations of the profoundest reason and the most radiant beauty, which are only accessible to our reason in their most elementary forms—it is this knowledge and this emotion that constitute the truly religious attitude."

Spinoza's is a possible, Todorovian God: a fantastic rather than fantasy-based God, a God that may dwarf our understanding and command our respect but a God that does not contradict the hard-won knowledge scientists have garnered of a massive, magnificent, and impersonal universe. Spinoza, though a great advocate of freedom of the press and freedom to worship (his *Theologico-Political Treatise* is said to have influenced the United States' founding fathers), did not believe in free will. He thought it was an illusion, the result of our parochial ignorance of the higher state of affairs—one described by the eternal laws of nature, one outside of time, one not open to divine intervention. Spinoza argued for the deterministic world Einstein defended in his letter to Max Born when he wrote, "He [God] does not play dice." Freedom for Spinoza was equivalent more to the growth of knowledge than the real exercise of free choice.

The God of Spinoza, Einstein, and my father contrasts greatly with the image of an all-too-human celestial tyrant demanding human obeisance, the Judeo-Christian-Islamic entity whose ire takes the form of plagues and bad weather, who is not averse to speaking to mean-spirited mujahideen, corrupt politicians, and money-grubbing televangelists. This alternative sort of God—defined not just as the observable universe but also, perhaps, as what we can logically infer about it—is not dependent upon turning a blind eye to scientific evidence. In fact, Spinoza suggested that, just as Protestants had correctly dispensed with the corrupt Roman clergy in order to study the Bible on their own, so we may dispense with the Bible to study the eternal verities of logic, science, and our own infinity-imagining minds. The supreme being, if any, is not only around us, and beyond us, but inside us. Books may be tools and deep spirits point the way, but the ultimate ground of reality, insofar as it is available at all, is democratically available through inquiry and knowledge.

But it is not just creationists who are wrong. Even most scientists, while they may know that life doesn't violate the second law because it is an open system, are not aware that life belongs to a natural class of complex systems that actively produce more entropy than would be the case without them. This is understandable; nonequilibrium thermodynamics—the science that studies real (as opposed to computer-generated) complex systems—is relatively new. But those who believe thermodynamics' mandate toward atomic disorder and evolution's tendency toward increasing complexity are at odds with each other are incorrect—as are those who believe that evolution has no progressive tendency.

Complexity and the second law go hand in hand. The formation of organized systems such as hurricanes Rita and Katrina, for example, are best understood as manifestations of the second law, their miles of devastation coincidental rather than central to the law. Life also depends—integrally—on the second law. Indeed, that law can be viewed as both essential and an ever-present imminent threat to our lives.

The basic equation for the second law is amazingly simple: heat transferred/absolute temperature = entropy. The trouble is that the early nonmathematical definitions described molecular behavior before that behavior was understood, long before quantum mechanics. Thus, although "entropy is disorder" was acceptable at the time, it was mistaken. This never impeded the scientific progress of thermodynamics because it was based on mathematical relationships, not on descriptions. A shining example of the practical success of thermodynamics is German chemist Fritz Haber, who, in the early 1900s, produced inexpensive ammonia using nitrogen from the air and hydrogen gas. This has enabled heavily populated countries to make ammonia-based fertilizers cheaply, staving off starvation on a global scale. It also raises the stakes of the human game, because there are far more energy- and food-dependent humans upon the surface of the limited Earth—or rather, within the Gaian system—than there were before the start of industrial agriculture.

Perhaps the enormous success of thermodynamics, in both academic theory and industrial production, caused experts to ignore simple descriptions of what the second law means over the past century. That is changing. Publications in the *Journal of Chemical Education* by Occidental College professor emeritus Frank L. Lambert have drawn attention to the century-old nonscientific mantra, entropy-as-disorder, altering the modern textbook landscape. By May 2005 authors of twelve out of thirteen new editions of US general chemistry textbooks had deleted their previous identification of *entropy* as "disorder." The new texts describe the second law as being fundamentally a matter of energy dispersal. If energy is not hindered, it spreads out.

Like fire.

This natural tendency to delocalize, moreover, can lead both to more as well as to less organized structures. Bacteria, you, and Gaia are examples of the former.

Nobel Laureate Steven Weinberg has written that science moves by the enunciation of "simple, impersonal principles." The second law is such a principle. Energy disperses if it is not constrained. Because the measure of the dispersal process, entropy, is an abstract ratio—q/T (heat/temperature)—it is not so easily grasped as quantities such as volume, pressure, or temperature.

A striking example of the confusion about entropy that is due to its being a ratio is the phrase *low entropy.* The entropy of any perfectly crystalline substance at zero temperature is zero. A substance at a moderate temperature whose molecules might move relatively slowly (and thus have a quite small amount of thermal energy, "heat") has a "low entropy" (quite small q/moderate T). But if that same substance has a large amount of energy within it and its temperature is huge, say thousands of degrees, we can see that the entropy would also represent a "low-entropy" state—a large energy divided by a very much larger temperature. *Low entropy* can thus refer to radically different situations.

The most basic example of the second law is neither difficult nor complex. Place a slightly warmer piece of iron on another piece of iron: the "heat energy" flows to the cooler iron until the two become exactly the same temperature. Technically, the "heat energy that flows" is actually the vibrational energy of atoms dancing in place in the metal that, on average, are moving faster in the warmer iron bar than in the cooler. At the surfaces of contact, the vibrations of the warmer bar interact with slightly slower vibrations in the cooler bar, whereupon over time the surplus energy of the warmer bar disperses. It spreads out so the vibrations of atoms held in place are at the same energy levels in both bars.

Quite clearly, no "disorder" or "order" is involved here! The atoms and molecules of all substances above absolute zero temperature are incomprehensibly disorderly, capable of being in any one of a truly gigantic number of different arrangements of their energy. The crucial point is that energy dispersal or spreading out is at the heart of any process involving thermodynamic entropy change—whether in iron bars or complex chemical reactions.

Such second-law-fostered energy flow toward dispersal is central to the process of life. In metabolism, as well as fires, oxygen reacts with relatively greater-energy-content substances to produce lower-energy-content products, carbon dioxide and water, CO_2 and H_2O, plus heat. This is energy spreading out—from localized greater-energy compounds to lesser-energy compounds plus heat. But then why don't all our many higher-energy biochemicals catch fire immediately—why doesn't everything "flammable" in our house do so, too? The reason is simple. Energy is required to break bonds. That is why a small fire with readily flammable paper or wood shavings is necessary to begin a large fire. The little fire furnishes activation energy, E_a, to break a few cellulose bonds for combustion to start. Then the energy given as heat when the first few lesser-energy bonds of CO_2 and H_2O form is large enough to continue breaking nearby cellulose bonds, produce more carbon dioxide and water, and spread out the great amount of energy we sense as a roaring fire.

Similarly, but far more slowly, the digestive processes within our body oxidize the portions of our food that are not utilized in forming other vital replacement fragments. We capture some of that second-law energy being dispersed and store it or use it to make higher-energy biochemicals for muscle movement and thought. A sort of natural cyclical fire, we take in oxygen and foodstuffs to act, and store some energy so that we can find more food to act and store, so that we can . . .

Ecologist Eric D. Schneider developed a formulation of the second law similar to Lambert's emphasis on dispersal—one that also largely dispenses with the confusing term *entropy.* A former EPA district chief who became disenchanted with crude official measures of toxicity (the government agency required a majority of hardy fish to die before water was tagged toxic), Schneider spent decades trying to measure ecosystem health. His studies led him to thermodynamics and the second law. In Schneider's reformulation the second law describes a process in which gradients—differences across a distance—are naturally reduced. Schneider points out that as these differences in temperature and pressure tend naturally to resolve, they produce energy flows—including energy flows central to organized, as opposed to disordered, systems. Thus cyclical chemical reactions and ordered regimes of heat flow called convection cells that range in size from microscopic to atmospheric are organized by the second law. Heat flow spontaneously produces quite complex structures as it spreads energy. As George Mason University molecular biophysicist Harold Morowitz says, "matter cycles, energy flows." We see such material cycling and energy dissipation in hurricanes and life. Hurricanes such as the Gulf Coast's Katrina and Rita do not develop over deserts or polar regions where large energy gradients between surface and upper atmospheres are rare and the essential energy-transfer agent of moisture is absent. It is the enormous area plus moisture between a very warm ocean and cold upper air streams—the gradient—that permits the spontaneous whirling complexity of hurricanes. Not only do Schneider and others surmise that life may have originated similarly—as a gradient-reducing system—but satellite measurements of ecosystems show that complex ecosystems reduce energy gradients better than simpler ones, and all life tends to disperse energy better—more quickly, or more sustainably, or both—than nonliving systems. The secret of life may be as disappointingly simple as the secret of a magic trick: not a concealed string or bit of wax but the second law. Throughout the universe the second law favors atomic chaos-producing, gradient-breaking energy dispersers. Despite the bias we have from being a shining example of the system under question, life may be just another naturally occurring matter-cycling energy-spreading system.

The fundamental action of the second law is energy dispersal. In a powerful hurricane this process obliterates all but the sturdiest structures. Although on a different scale, life is similarly inured. An organism uses some of the energy that might otherwise harm it to make complex, sturdy biochemicals that are protected from second-law action by energies of activation, E_a. While energy flow channeled by reproduction and evolution preserves life's huge cyclical system, *death* is the name we give the inevitable disruption of a specific part of life's intricate network. Life's trick has been to preserve, expand, destroy, and replace connections in its huge, four-thousand-million-year-old modulation of energy flow. Crafty life not only obeys the second law, but takes advantage of it, diverting energy to make the sturdy biochemicals that resist typical second-law change.

Down, down, down

into that burning ring of fire . . . up, up, up, with the flames getting higher," croons Eric Burdon, from the Animals, in a classic rock song written by June Cash and Merle Kilgore. The Gaian macroorganism, like all complex thermodynamic systems, feeds on energy as it cycles its matter.

Part of what life has stored during the course of its energy machinations under the sun is oil, coal, and natural gas in the ground. We, of course, have come, with our precocious intelligence that has so much problematic promise, to plunder these deposits with their considerable potential energy. We of the Holocene are now in a position where all signs point to a general reduction in Gaia's ability to perform at maximum efficiency. Remember, thermodynamic systems disperse energy, mostly as heat, which tends to keep them cool, both figuratively and literally; but like machines, they can overheat and lose their functionality. Living "machines," of course, differ from nonliving machines in that they are instrumental only in maintaining their own operations. They are "for" making more of themselves, not something else. And while entropy production drives them, the energy running through them can also threaten their energy-using efficacy.

Successful evolutionary forms have navigated the rough waters, steering between the rock of insufficient energy and the hard place of overusing the energy sources on which they depend. The loss of hair in the evolution of humanity itself can be traced to the superior ability of naked skin to evaporate sweat, a key ability in our ancestors who needed to sweat to keep up on the hunt. Sweating without interference from too much hair helped keep our ancestors from overheating, and gave them the advantage we still enjoy as the fastest long-distance runners, the best hunters this planet has seen. Gerontologist Walter Bortz further speculates that running, by enhancing blood circulation, and therefore oxygen to the brain, is part of the story of the rise of cranial intelligence.

But our species pride—increasingly good energy users from the days of the hunt through to agriculture and industrial technology—is the planet's peril. Global warming—it is hotter now than it has been since some 55 million years ago, during the Eocene, a phenomenon that may have been caused by natural methane ice crystals (methyl clathrates) melting—is a symptom of thermodynamic impairment. Earth's surface has been materially injured by population and technological growth. It is going into a "ring of fire"—heating up at the surface in a way that may bring on sudden chills as well as a fever and other signs of disrupted global climatic functioning, such as increased storm systems. Systems thinker and philosopher Gregory Bateson described the wild wavering of systems as "schismogenesis"—a splitting that leads to a new regime. Gaia, its subsystems grappling with the effects of rampant human growth and energy use, may be about to enter a new climatic regime.

Although the all's-fair-in-business-and-war departments of some major oil corporations, such as Exxon, have been working hotly behind the scenes to dismiss evidence of global warming, the majority of scientists agree that the recent and continuing rise in global mean temperature can be traced mostly to CO_2 emissions from human industry and cars. According to an Air America broadcast of February 2, 2007, the American Enterprise Institute, backed by Exxon, was offering ten thousand dollars for scientists to dispute the 2007 United Nations Intergovernmental Panel on Climate Change, which predicted temperature rises of two to eleven and a half degrees Fahrenheit by the year 2100. The destruction of forests to make room for crops may also play a role. As mentioned, forests cool the planet by the process known as evapotranspiration, which, like sweat on the skin of a marathon runner, is a good thing.

Gaia, Lovelock says, likes it cool. But we have been warming the biosphere. The result is a collision course, one in which we are decidedly *not* the favorites. To reiterate, the Gaia hypothesis states that the chemistry and temperature of the atmosphere, oceans, and sediments—the surface of the Earth—are under active control by the biota, the organic surface of the Earth. Oxygen accounts for about one-fifth the atmosphere; the mean temperature of the lower atmosphere is about twenty-two degrees Celsius; and the pH of the planetary surface averages just over 8.0. Chemical evidence from the fossil record shows that such anomalies can persist for many millions of years. There is nothing vitalistic about their active control by life: Life, like other thermodynamic systems, shows its complexity only within certain parameters, such as temperature boundaries and other constraints. The more energy is available to life within these constraints, the more life does its thing. Increased incident temperature leads to increased heat and entropy production. Coolness is maintained naturally.

The Gaia hypothesis gives an answer to a mystery that has been gnawing at scientists for decades: the so-called faint young sun paradox. Astrophysical theories of the birth and death of stars uniformly suggest that the sun began five billion years ago, smaller than it is now; it progressively increased in size and luminosity by some 30 percent or more. Yet fossil evidence shows that liquid water, including rivers and lakes, existed at the Earth's surface for the last three billion years. How did the biosphere cool itself to make up for the sun's growing luminosity and the increased radiation reaching the Earth?

The traditional explanation for such temperature control is luck. From the Gaian perspective, however, life actively participates in climate control. Many animals regulate their temperature. Honeybees even act together to modulate the temperature of the hive outside their bodies proper. Might not the Earth, too, regulate its temperature, chemical composition, and environment in general not through any outside agency but as the aggregate result of the organisms busy within it? These organisms, growing only within certain temperature parameters, have the net effect of keeping the global environment cool despite the increasing overheating threat from a sun that, in keeping with the series of nuclear reactions that defines star development, becomes more luminous over time. Although life seems to "know" how to react as a whole to maintain the conditions of its existence, the same may be said of the air molecules swirling within a tornado—and no one would claim their behavior to be impossible because a tornado requires inexplicably coordinated action on the level of the whole.

The global cooling action for which there is good fossil evidence is now being compromised by rampant human population growth, removal of forests, and the release of greenhouse gases in the atmosphere. Not only is global warming real, Loveock argues in *Revenge of Gaia,* but the stakes are so great that we must as a global civilization move immediately to relatively clean and unlimited nuclear fuel if we are to survive. Although even mainstream politicians (reared to moralistically enrich themselves as they tell their constituents what they want to hear) are beginning to advocate action to curb greenhouse gases and global destabilization, time is running out. One of the controversial short-term technological fixes that have been suggested to stall warming while we get our global environmental act together is to spread a haze of sulfate particles in the upper atmosphere; reflective aerosols could, for example, be added to jet fuel, in effect applying a global sunscreen. Futuristic windmill-like machines with biomorphic titanium-dioxide-daubed blades have been designed that could scrub carbon from the air. Atomized seawater might be used to seed clouds over the ocean, increasing the whiteness and reflectivity of the atmosphere. Giant mirrors might even be deployed in space. Such dialysis-like stopgap measures will not in themselves solve the problem. They will only buy us some time.

It is quite possible that we are on the verge of the biggest environmental crisis in perhaps a million years. As we know from written accounts from various cultures, there is in our history the persistent myth of a flood—most famously Noah's in the Bible but also

Atlantis. As the International Panel on Climate Change has reported, and although it does not sound like much on paper, a five-degree Celsius temperature increase by 2020 will literally change the world. By then the intolerable European summer of 2003 will seem downright cool. Floods will submerge what is now dry land, and other lands will turn to desert. Volcanoes the size of Tambora, the largest historical eruption on this planet—which occurred in Indonesia in 1815—will be common. Humankind, or what's left of it, will migrate en masse to cooler climes—such as newly exposed land surface around the North Pole, currently covered by ice.

We are somewhat prejudiced because we think the term *ice age* sounds draconian. However, planetary life as a thermodynamic system is healthy insofar as it produces heat as entropy, keeping itself cool. It seems to do this in part by, in the short term, producing reflective clouds and in part by, in the long term, burying carbon in the ground as organisms, after taking carbon dioxide into their bodies during photosynthesis, pile up to become coal, limestone, oil, and other carbon-rich life-deposited substances. This effective long-term carbon burial mechanism has been upset by our discovery of buried energy sources, making the Earth system more active, but also polluting the global surface environment to the point that Gaia less effectively performs its thermodynamic function. In fact, the cold temperate areas of the ocean have much more life in them than tropical areas, where the water is a crystalline but relatively lifeless blue. To restore health to the global ecosystem means effectively counteracting global warming. This may, however, be impossible. We may be beyond the point of no return. Alternative energy isn't going to cut it, Lovelock says. While geothermal energy will work well in Iceland, solar power will not work for cloud-covered England, and wind power is a quixotic fantasy that will never replace the quantity of electric power that runs a modern city. Organic food can't save the megalopolis, which depends on, is addicted to, plentiful energy. And biofuels fall into the category of foolish short-term fix because petroleum-based fossil fuels and farm equipment are used to grow the plants that would save us from dependence upon energy! We already use 40 percent of Earth's surface to feed humans. And a car uses forty times more energy than a person does. What happens when India and China become as automotively urbanized as Los Angeles and New Jersey?

We are between a technological rock and a climatic hard place. I mentioned earlier Garret Hardin's tragedy of the commons: the phenomenon in which individual organisms, by pursuing their selfish interests, ruin it for everyone, including themselves. I encountered a mundane example of this recently on a visit to the local library: The headphones, which had previously been set out for use with each computer, had all been stolen, and the library wasn't budgeted to replace them. Similarly with regard to energy and the environment: The relentless pursuit, come hell or high water, of diminishing energy resources will in the end deprive the global community of affordable oil and gas even as it brings on both hell in the form of war and high water in the form of global climate change. The only possible answer to the gargantuan ecological mess we've gotten ourselves into is that old environmentalist nightmare, nuclear power. Hydropower works for Norway, and natural gas for Russia. Germany pioneered wind turbines, but they are operational only 16 percent of the time. For the vast megalopoli, the global human technological monster of which we are a part, and which would shrivel up and die without its huge energy influx, stronger medicine is needed. That medicine, in Lovelock's estimation, is nuclear energy.

While it elicits a knee-jerk response of environmentally correct naysaying, based not entirely on the facts, only nuclear energy can save the planet from the climate catastrophe that will otherwise—through a combination of starvation, habitat disruption, political unrest, and so on—whittle the global human population down to perhaps one-sixth of its current crowds. The report of the latest Intergovernmental Panel on Climate Change, a conservative consensus of some two thousand of the world's leading climate scientists, predicts a global mean temperature rise this century of between 2 and 4.5 degrees Celsius (3.6 and 8.1 degrees Fahrenheit). This level of rise is projected to lead to deadly heat waves, increased numbers of more powerful tropical storms, and sea levels rising from inches to feet. Such a temperature increase will also shift rainfall patterns, possibly leading to permanent drought in agricultural breadbaskets such as the US Midwest. Disruption in the alimentary conduits of the human superorganism will foment mass starvation, migration, and resource war. If urbanized coastal areas are flooded, the situation of course will be exacerbated.

It is too late to go hang gliding to the garden to gather food for the family supper. We are entering an age of the politician's proverbial "tough choices." In fact, France already gets 80 percent of its energy from nuclear fission, the highest percentage for any country; half of the energy needs of Connecticut are met by its two nuclear reactors. Nuclear power is not necessarily obvious; Vermont, one of the most pristinely natural states left in the United States, generated 72 percent of its energy in 2005 from nuclear power, the highest rate in the country. While China, India, Iran, Pakistan, and South Africa either have, are building, or are planning to build nuclear plants, the United States is the biggest user of nuclear energy, already the second largest source of energy in the country. A 1990 National Cancer Institute investigation of possible links between cancer and nearby nuclear facilities found an increase of deaths neither from cancer nor from childhood leukemia. The Nuclear Energy Institutue points out that one would have to live near a nuclear installation for two thousand years to receive the radiation equivalent of a single diagnostic X-ray. And the relationship between nuclear energy and terrorism is weaker than you might think. Apart from the fact that enriched uranium lasts only eighty to one hundred years, a nuclear power station is not directly related to construction of nuclear weapons. You're no more likely to use a nuclear power station to build a nuke, according to Lovelock, than you'd be to use coal fire stations to build conventional weapons. While the dangers of nuclear weapons are truly horrific, the dangers of nuclear energy may be exaggerated. In fact, a Swiss study shows that, of all modern methods of providing power to cities, deaths per terawatt of energy are lowest—by a factor of ten—using nuclear power. A nuclear energy solution obviously is not perfect or ideal—but since when in life has anything ever been?

The mass dependence
on energy of the human superorganism will not permit the six-plus-billion of us in the Holocene to survive on oil, coal, and gas alone. In a striking analogy, Lovelock compares our situation to that of a transatlantic jet full of environmentalists who realize, midair, that their plane does not have sufficient jet fuel to land; with seeming prudence, they decide to conserve—"let's go like a hang glider." Of course, this will not work. Industry-backed global warming denialists say there is no problem; environmentalists say there is but present an insufficient solution. Even a change of heart among oil company propagandists, a massive global effort funded by the fiat money of many governments, and wholesale acceptance of solar power may well not be able to support the human race at its present-day energy levels—which are disproportionately represented by members of the United States, whose standard of living is coveted by far more populous India and China. The extreme noxiousness to the environmentalist of a nuclear technological fix needs to be viewed in the context of the severe deprivations that will arise if heating, urban electricity, and global food production—presently dependent on fossil fuel—are curtailed.

Our tendency to deny both the phenomenon and the drastic measures needed to battle it is reminiscent of heavy smoker who, upon being told he has cancer, reaches for a cigarette to relieve the stress. We use what Freud called "kettle logic." Based on a German joke, this is the sort of "logic" that appears in dreams, which, being unconscious, are not inhibited by rational prohibitions against contradiction. A man is reprimanded for returning a damaged kettle. His response is that he never borrowed the kettle. He adds that he returned it undamaged. Finally he says that the kettle was already broken when he borrowed it. Humanity's grasp of global warming may be similar. The industrial apologists say there is no global warming—we never borrowed the kettle. Or that there is global warming, but it's not humans' fault—the kettle was already damaged. And the environmentalists think we can return the kettle undamaged—all we have to do is not emit quite so much CO_2, that is, implement some prudent conservation measures, and everything will be okay. But a Band-Aid or even aluminum foil placed over the gaping hole will not make the kettle whistle. We cannot glide safely in a Boeing 747 to a buoyant ocean runway. We must grapple with the quite real possibility of a climatic crash. If the reality is something we can accept and wish to avoid, then drastic measures may be called for if we don't want a wholesale return to the hothouse conditions of the Eocene. And even if we do—even if the requisite geopolitical will materializes to make a full-fledged global effort to replace our energy sources—it may already be too late.

But there are possibilities. Paleontologist Jessica Whiteside, a veteran of the noted Lamont-Doherty Earth Observatory in Palisades, New York, has a plan to save the planet more drastic than Lovelock's politically problematic embrace of nuclear energy. Desperate situations require desperate measures. After all, this is not a kettle that is damaged and interfering with teatime, but the cybernetic Earth system, interfering with human habitability of our planetary home. As Shakespeare's Claudius says in *Hamlet, Prince of Denmark,* "Diseases desperate grown, by desperate appliance are relieved, or not at all." Whiteside, now in the Geological Sciences Department at Brown University, has proposed, apparently not in jest, a fascinating, if at least superficially terrifying, solution to the problem of global warming. Her idea? Forced weathering of calcium and magnesium silicates of basalts would react with CO_2 to produce limestone if great quantities of air could be forced into it. The proposed process, carbon sequestration, essentially reverses the decades-long process of oil and gas companies using every means at their disposal to retrieve carbon-rich fuels from underground; in carbon sequestration, CO_2 from the air is put back into the ground. In 2005 the Big Sky Partnership, a collaboration among the Department of Energy, several universities, and national laboratories was established to explore ways of carbon sequestration. The name Big Sky arises from the fact that the states of Montana, Idaho, Washington, Wyoming and South Dakota currently show the most promise for storing carbon. This is because they contain an estimated 53,000 square miles of basalt, deep, porous, volcanically formed rock whose chemical composition makes it a leading candidate for carbon sequestration. Most thinking on the subject so far has focused on long-term projects of drilling or pumping CO_2 into the basalt. In a creative leap Whiteside has offered a rather amazing and perhaps quite serviceable idea along these lines for stopping the greenhouse effect not slowly but dead in its tracks. It is, however, dependent upon what undoubtedly will be a politically contentious solution: exploding nuclear bombs within basalt deep underground to profoundly fracture the rock, thereby providing more reactive surface area to neutralize the CO_2. Contentious, to be sure. But as they say about death, it's not that bad considering the alternative. Life's habit since its tumultuous beginnings on the Hadean Earth four billion years ago has been to raise the stakes of its crazy metabolic and always quasi-technological game: Nuking (parts of) the planet to save it sounds potentially disastrous but, as an example of triage potentially saving humanity from losing its civilization, if not its very existence, this radical proposal seems eminently worthy of serious consideration. A quirky friend even ironically suggests a possible logo—a rainbow connecting two mushroom clouds above the motto, "We did it together"—for this extreme climate project.

And so we move from the tactic of wholesale adoption of nuclear fission, building (relatively) clean and safe nuclear power plants, to nuclear explosions. Tactical nuclear explosions set off in particular geological areas could instantaneously suck huge amounts of offending carbon dioxide out of the atmosphere. Basalt rock, which contains magnesium and calcium, would, in the presence of a nuclear explosion and the accompanying atmospheric pressure gradient, experience a wicked tunneling of air into its nooks and crannies. There the carbon dioxide in the air would react to form limestone, not over eons but in a geological blink of the eye. Of course, the political will to accomplish this program would likely meet with even more resistance from would-be environmentalists than Lovelock's criticized program. The problem is that even if the time frame of the average environmentalist is longer and wiser than that of the average corporation's next quarterly report, it is not long enough. Driving down the ecological freeway we risk not just running out of gas—worldwide fossil fuel depletion—but also slamming into the "unseen" wall of climate collapse as we blithely use resources that have so far been provided us for free, courtesy of the metabolic activities of other organisms.

I attended a lecture by James Lovelock on October 14, 2006, at the American Museum of Natural History. Lovelock, an inventor and visionary rather than a professor, is known to spend months preparing for a lecture. As luck, or Gaia, would have it, I was sitting between him and the aisle before he gave his presentation. Nervous before going on, he asked to switch places with me so his access to the stage would be easier. I squeezed his arm and told him he would be great, then settled in my new seat and scribbled furiously to record some of what I have reported above. Lovelock's commentary on the necessity of radical measures in treating a feverish Earth broke the ice as far as suggesting the even more radical measures suggested by Professor Whiteside. At the end of the lecture I approached him and apprised him, in abbreviated fashion, of her idea.

"I think I understand what she's saying," he said, nodding. "Yes." He smiled. "I think they'll try that. They'll try lots of things."

In February 2007 the *Washington Post* Foreign Service reported that British billionaire Richard Branson, the handsome owner of Virgin Atlantic airlines and friend of the Rolling Stones, with former vice president Al Gore at his side, was offering a twenty-five-million-dollar prize for anyone who could "come up with a way to blunt global climate change by removing at least a billion tons of carbon dioxide a year from the Earth's atmosphere. . . . The first five million would be paid up front, and the remainder of the money would be paid only after the program had worked successfully for ten years." The judges included James E. Hansen of NASA's Goddard Institute for Space Studies, James Lovelock, Sir Crispin Tickell, and Australian conservationist Tim Flannery. There was only one glitch: "The winner of the award must devise a plan to remove greenhouse gases from the atmosphere without creating adverse effects." Upon hearing this (and thinking I might make a quick million by helping to write the proposal), I spoke with Whiteside. But she seemed more concerned with looking for glutaraldehyde to use as a fixative for a jar of dying termites whose teeming hindguts, full of many species of microbes, she was keen on displaying to her students during her next fossil record class as an example of life's unsuspected richness.

What are the chances?

A few summers ago I was off the coast of New Hampshire on an island frequented in olden times by pirates and still perhaps containing buried treasure. It was night, I was inebriated, and I suddenly found myself in the presence of a clearly desirable blond woman. (She had said she missed my lecture because she just wasn't interested in that science stuff. I told her I could totally understand what she was saying.) Then there was a rumble of thunder and she smiled at

POKER PROBABILITIES: THE CHANCES OF BEING DEALT

One pair—drawing two of the same kind (for example, two kings) in a hand of straight poker ("five-card draw"), nothing wild: 1 in 2.333

Straight—five cards in sequence with mixed suits, for example 4♣,5♦,6♥,7♠,8♦: 1 in 250

Flush—five cards of the same suit not in sequence, for example, K♥,7♥,2♥,3♥,9♥: 1 in 500

Full house*—Three of a kind with a pair, for example, 7,7,7,A,A: 1 in 700

Four of a kind—for example, four queens; the remaining card doesn't matter: 1 in 4,000

Straight flush—five cards in sequence all of the same suit, for example, 6♥,7♥,8♥,9♥,10♥: 1 in 65,000

Royal flush—Straight flush of highest cards, for example, 10♠,J♠,Q♠,K♠,A♠: 1 in 650,000

You reading this—Unknown: It could be 1 in 1 (if this is the only reality) or 1 in infinity (if there are an infinite number of other realities you're not aware of in which you don't read this!)

***Coincidentally (?), just as I finished writing this I heard the announcer on the radio say the jazz composition we had been listening to was "Full House" by guitarist Wes Montgomery, with Johnny Griffin on tenor sax, Paul Chambers on bass, Jimmy Cobb on drums, and Wynton Kelly on piano.**

me. "Can you believe this?" she said. "You, me—us here, in this place, on this planet, with these stars, on this little island in the ocean in the middle of nowhere. What are the chances?" She gestured. I looked out at the rocks ringing the island. Suddenly lightning lit up the skies as if to illuminate her point. What are the chances?

Nobody knows. We don't know because this Earth, this life, our status as a bipedal mammal with a particular species, ethnic, family, and individual history is the only example we have. On the one hand, we may be a foregone conclusion; it has yet to be proven that anything could have been the slightest bit different. On the other, this universe may be one of an infinite number. One over an infinite number. That's a lot of universes. Us as an infinitesimal fraction.

But here we are.

Life is fire-like
in its growth, in its energy capture and deployment, in what Nietzsche called its will-to-power. Vernadsky referred to it as "green fire," the transmutation of sunlight into green matter and the organisms—like us—that eat them. Where will such fire end?

In one of my stranger speculations—which I characteristically presented as science—I suggested that, even if the universe wasn't yet alive, the biosphere, on the brink of reproduction, was about to enter a rapid-growth phase; unwittingly using human technology to make more of itself, it would spread like wildfire, turning the entire universe alive. Plant geneticist Wes Jackson warns that the dangers of technology, especially of biotechnology, may be moving forward far too fast for the public good. Jackson claims that the Gaia hypothesis—that the Earth is an organism, or at least a physiological entity—is not just poetry, but bad poetry. To prove this, he points out that all organisms have offspring. For the Earth to be an organism, there would have to be planetary offspring. The Earth has none. Therefore the Earth is not an organism.

The problem with this seeming logic is that it argues on the basis of what has been, rather than what may be. It is not difficult to imagine casinos in enclosed ecosystems on Mars that recycle their chemical constituents in the manner of Earth's life. While Wes Jackson's point that we don't know what Earth is, any more than we know what God is, is well taken, his attempt to prove that Earth's surface cannot be a true organism because it has no offspring suggests the exact opposite: that Earth, availing itself naturally of human-produced technology, is evolving the potential to make "baby Earths" in the sense of new thermodynamically closed recycling ecosystems.

And if Earth's life made it, in a sustainable way, to extraterrestrial shores, that would represent, to my way of thinking, de facto propagation of Earth as a bona fide organism. The argument that Earth is not an organism because it has not reproduced is null and void because ecologists and space voyage designers actively attempt to understand the mechanics of setting up a completely recycling ecosystem. This ecosystem has been in continuous operation on this planet since the early evolution of the starship Earth. However much we, in our penchant for self-involved cosmogonies, insist that only our technical civilization would suffice to bring, Noah-like, the sprigs of our garden to the stars, it seems likely that evolution's root tendencies of increasing energy storage and deployment, memory, strength, mobility, and intelligence may be more responsible for the possible reproduction of the Earth system. In the end, with or without us, reproduction is natural. It has nothing specifically to do with humanity. Cells reproduce and groups of cells reproduce; symbiotic collectives reproduce; viruses and even technological gadgets and clever phrases seem to do it. Why should we assume the Earth itself cannot reproduce? According to creative writer, evolutionist, and Darwin's New Zealand sheep-farming contemporary Samuel Butler, we are not simply the children of our parents. We are also "children of the plough, the spade, and the ship; we are children of the extended liberty and knowledge which the printing press has diffused." We are not just human but "tool-kits" fashioned for itself by "a piece of very clever slime" over very long experience. And if our reproduction is not so simply biological as we may imagine, the same may be true of the biosphere:

> It is said by some with whom I have conversed upon this subject that the machines can never be developed into animate

> or quasi-animate existences, inasmuch as they have no reproductive system, nor seem ever likely to possess one. If this be taken to mean that they cannot marry, and that we are never likely to see a fertile union between two vapour-engines with the young ones playing about the door of the shed . . . I will readily grant it. But the objection is not a very profound one. No one expects that all the features of the now existing organisations will be absolutely repeated in an entirely new class of life. The reproductive system of animals differs widely from that of plants, but both are reproductive systems. Has nature exhausted her phases of this power? . . . What is a reproductive system, if it be not a system for reproduction?

Although I believe the production of a single successful recycling environment would represent de facto reproduction of the biosphere—thereby proving Lovelock's original assertion that the biosphere is an organism—we need not go that far. In my *Biospheres* I argued that reproducing Gaia is part of a "technobiontic" wave in which biological individuality crops up at ever-more-inclusive levels. If bacteria use magnetite to orient themselves and birds distribute phosphorus on a global level, it is hard to argue that we humans with our GPS systems and FedEx delivery systems are the only technological beings. And if technology of some sort is ancient, then the inclusion of it in biospheric means of reproduction is not a cogent argument against the fertile, growing biosphere. Moreover, it occurred to me, after discussions with my mother, biologist Lynn Margulis, that the sort of protective enclosures being developed in the Arizona desert were similar to other forms invented by seed-bearing life at a variety of scales. The more I thought about it, the more Biosphere 2, Clair Folsome's flasks, and other such designs looked like planetary spores. This I considered unsurprising, since so many kinds of life had already happened upon the stratagem of building hardy forms capable, under propitous conditions, of rapid growth.

To understand the assertion that biosphere formation recapitulates ancient means of organic survival, let us consider some examples more pedestrian than the Earth as a living whole. The endospores of the bacterium *Clostridium* are propagules. So are the seeds of plants and eggs of animals. Hibernating in caves is the propagule-like behavior engaged in by some bear species. The tardigrade or "water bear," a pond invertebrate that through a scanning electron microscope looks amazingly like a bear, exemplifies the effectiveness of propagule strategies. When food becomes scarce, or when a tardigrade incurs bodily damage, it forms a "cyst," a thick-walled enclosure in which the animal contracts and its organs deteriorate. Yet the tardigrade is not dead; it can regenerate to live again. Tardigrades also periodically dry out, metamorphosing into immobile, barrel-shaped forms called tuns—the name for wine casks, which describes what they look like. As tuns, water bears can survive a very long time—some people say a hundred years. In their various protective enclosures, tardigrades can withstand temperatures ranging from 151 to –270 degrees Celsius (a temperature that freezes into liquids atmospheric gases such as nitrogen and oxygen). And they can tolerate an X-ray dose of 570,000 roentgens (which would kill a person a thousand times over). Not surprisingly, considering their ability to form propagules, tardigrade species inhabit both the arctic and the tropic zones, as well as places in between. In the same way, human efforts to assemble organisms into joint recycling enclaves smaller than the biosphere as a whole (arguably the only complete ecosystem we know) suggest that Earth is *already* producing propagules preparatory to propagation. Gaia may not have reproduced, but the weight of evidence of Folsome's flasks, ecospheres, recycling fish-and-algae designs, the Soviet *Bios* projects, and Biosphere 2, even if it failed, suggest she is budding.

The third law of thermodynamics basically says that you can't reach absolute zero. At absolute zero, zero Kelvin, all particle movement would stop. The fact that there is always, even in deep space, some background motion and temperature above zero suggests that the universe is, in a way, alive. Certain life-forms, including some bacteria, can be frozen to near absolute zero; when rewarmed they revive. This ability of life to hole up, to maintain its structure and wait for a better time, when available energy will bring it back to life, has given some hope of immortality. Red Sox baseball great Ted Williams, for example, paid a hefty sum of $136,000 to have his head removed and placed in liquid nitrogen in cryosuspension in Scottsdale, Arizona; he is among those who hope that future technology will be able to resuscitate their illustrious careers and successful lives, or preserve their biological essence until such time as medical science discovers the keys to true immortality. In Philip K. Dick's story "I Hope I Shall Arrive Soon," first published in *Playboy,* a cryogenically preserved passenger, en route to meet his wife on another planet, never definitively arrives. His daydreams of reunion with his mate, conflated with memories of their past moments, are modified by the ship's computer to entertain its frozen charge while on his long interstellar voyage. As mentioned, Lovelock was an early researcher on the forefront of cryogenics, the science of freezing life to preserve it—a great theme of science fiction, as it offers us egotists scientific hopes of cheating death, on the scientific basis that the essence of not only life but also consciousness adheres in the chemical structure, which can be accurately preserved. But I worry about these rich people who buy their way out of the graveyard, who attempt to give death the slip. I am afraid that they are experiencing the alternative. In other words, that they haven't died at all, but are just very, very cold. As in those dreams when you open your mouth to scream but can't, the cryogenically frozen may even be suffering but unable to call out audibly for help. I hope this isn't true, but I remember the words of the great Argentine writer Jorge Luis Borges: that nightmares are proof that Hell exists. I prefer to think of his opposite and apposite dictum, that libraries are something very close to Heaven.

The possible preservation of individuality by freezing demonstrates the key role energy plays in any commonsense understanding of life. Energy is the only life, as William Blake wrote. Life is not just latent structure but the metabolism, movement, and reproduction dependent upon energy flow. It is active, as the vernacular expression *to put life into* something, as well as the medically noted relationships between exercise and health, on the one hand, and exercise and happiness on the other, suggest. You do not have to be preserved as a genetic fragment in a freezer. If you're sleeping and hibernating, or just loafing, you're not as alive in the common understanding of the word as you are on the dance floor. Life is active. It transforms energy. Its heart is not that of an ice cube but of a raging fire. Nonetheless, it has the ability to wait things out, to bide its time by preserving its hardy structure. Although technological, living planetary propagules partake of an ancient process in which living forms—be they bacterial and fungal spores, plant seeds and pods, or animal forms such as tardigrade tuns and hibernating bears—enter a low- or no-growth phase, which allows them to endure uncertain times, increasing their chances of future survival.

There is no assurance

that reproducing biospheres, representing the flame-like spreading of fire-like life, will establish recycling enclaves in orbit, let alone the outer planets or other star systems. Perhaps the expanding universe, which astrophysicist Eric Chaisson suggests is a prerequisite for increasing complexity—because it ultimately gives complex forms a place to dump their entropic refuse—is also the cosmic precondition for life's loneliness. Even if we colonize the Moon and Mars and other nearby planets, we may fail in our attempts both to meet other life-forms and to spread to other star systems. It is possible that, natural as it may be, the spread of our technical civilization, as well as Gaia's reproduction via biospheres, will prove impossible due to the vast spaces between the stars and the expansion of the universe. That there is hope, as Kafka said, but not for us. The rates at which stars and galaxies are moving away from one another may prove prohibitive to space migration. We may expire on our grain of blue sand; the cosmic water droplet with its human microorganisms may dry up without so much as a peep. We may be victims of the great vast black space between stars, our cosmic isolation a counterpoint to the personal loneliness that French writer Maurice Blanchot calls "the essential solitude." The idea that Earth life has opened up some kind of rift in space-time, that we are important and full of cosmic promise despite our minuscule size and fleeting tenure compared to the life span of the cosmos, may be, well, wishful thinking. On the other hand, we are along for the greatest ride imaginable. However seemingly finite, life is connected to the universe, which may be infinite.

This idea—and a notion of reproduction ultimately far more grandiose than my talk of reproducing biospheres—has been seriously proffered by the lawyer, writer, and well-regarded amateur astronomer James Gardner. Gardner has suggested that intelligent life within the universe is not accidental but part of a universal process of cosmic reproduction in which evolving technological societies, such as our own, eventually learn to engineer new universes, perhaps by using nuclear lasers to build black holes. In this view, the purposefulness we notice in ourselves is not alien but integral to a universe that has random elements but is also on the way to producing new universes that, after sufficient time, will develop new intelligent civilizations. These intelligent civilizations, some of them, will spawn the astounding level of technology requisite to create new universes. As Arthur C. Clarke has said, a sufficiently advanced technology is indistinguishable from magic. Perhaps you remember one of the episodes from the original *Star Trek* television program in which the crew of the spacecraft *Enterprise,* upon being beamed down to a planet inhabited by giant humanoid aliens, realize that these same galactic Olympians long ago must have visited Earth's surface, wowing the Greek natives to the point that they devised their charming stories of imperfect gods. Such aliens, technological masters able to make interstellar voyages, might be sufficiently advanced to build new universes, perhaps creating black holes from orbiting laboratories.

My reproducing biosphere scenario, unlikely as it may be, pales in comparison with this reproducing universe hypothesis of Gardner. (His last name, by the way, indicates one of my personal favorite theories, which I call the Theory of the Vocational Signifier: that people's names, subtly or overtly, influence their life interest and choice of profession. In my theory, Gardner's last name—redolent with the intimations of fertility—is no more accidental than is our technological civilization as a natural outgrowth of a cosmic reproduction process in the theory he proposes.)

Gardner's notion has been taken seriously or, perhaps better put, entertained by NASA. Different universes theoretically could be "born" on the other side of black holes, those so-called singularities that are in many ways the opposite of the Big Bang, exploding into light and presence rather than being so massive that light disappears into them. These universes, moreover, could have variant physical constants and laws; they would be part of the invisible-to-us progeny of a previous "mother" universe that produced plentiful black holes, and via those black holes new universes, some of which gave rise to intelligent black-hole-producing civilizations.

Gardner's friend, physicist Lee Smolin, had already suggested that universes that come into being on the other side of black holes might do so with variant laws of nature, and that such variance in physical laws and constants of nature would lead to a natural selection of universes, with those universes whose laws are more conducive to forming more black holes tending to be better represented just as, other things being equal, rapid reproducers such as bacteria will be better represented than, say, elephants, let alone nonreproducing forms. Gardner's wrinkle to Smolin's cosmic natural selection idea is to suppose that life plays a key role. And not just any life but intelligent life, the rise of which may correlate with the technological know-how ultimately to commandeer the resources necessary for the Promethean project of universe production. A more Faustian role for humanity could scarcely be contemplated.

Papers have already been written suggesting the possibility that black holes could be produced, perhaps with nuclear lasers. Gardner projects humanity's technological progress to a theoretical future in which humans, or our progeny, produce new universes with new laws on the other side of black holes. Which suggests in turn that we ourselves did not arise de novo but were implicit from this black-hole-studded, "bio-friendly" universe from the start. Bizarre and brilliant as we gun-toting, plane-flying, nuclear-reactor-making humans may be, we manifest a version of a technical proficiency that is not out of step in universe whose cosmic reproduction process centers on the evolution of technology to the point that it can make new versions of itself through the cosmic orifice of black holes.

This is a flattering assessment. I gently poke fun at it because it seems both too specific and too vague. As when my father visited drug guru Timothy Leary in jail—Leary was keen to inform him that he was in constant communication with aliens, and wanted to impress upon his astronomer friend the need to visit them deep in space—there is an air of fantasy about this scenario. I am more prepared to believe that there are many, perhaps infinite, universes than that we are central to their production–reproduction. When my father asked Leary how we were supposed to get to space, Leary replied with cavalier confidence that that could be left to the engineers. So, too, despite the specification of nuclear lasers as a candidate for humanity's future world creation, I think it is safe to say that the divine technology of cosmos production by us or our descendants remains understandably murky. I am afraid that what Gardner's idea provides, through the back door, is once again a central place in the cosmos for humanity. That is the need it answers. We are no accident. Our burgeoning intelligence and technology, so transformative of our world, happen because we are central to creation, literally. Again we find ourselves, genetically this time, at the center. How convenient is that? Gardner applies metaphysical salve to our scientifically wounded collective ego. He makes us the sine qua non superstars of the universal process from which we were so recently demoted. This promotion of humanity's place in the universe rescues us from the Copernican backwaters that were threatening to eviscerate the last vestiges of our specialness. But psychological grandiosity and self-centeredness are, I maintain, hereditary concomitants of our behavioral and physiological requirements to maintain ourselves and our kind. Which means that, without denying its possibility, Gardner's felicitous fantasy seems improbable. Our understanding of the immensity of time and the physics of black holes, our development of the space shuttle and Prozac notwithstanding, we may *still* not be central to cosmic operations. As my father, God rest his soul—or the scientific energy equivalent thereof—said, "Extraordinary claims require extraordinary evidence."

Human beings

are not always so good at judging probabilities. Toss a fair coin one hundred times and quite often there will be a run of eight or nine heads in a row. Somewhere I heard a story of people who, wanting their arrangement of black and white tiles to be natural, found a way to randomly ascribe which tile would be put where. The final pattern looked so contrived, they had to remove it. True randomness may seem to us quite orderly, and vice versa. We tend to think we are special—extremely, even infinitely unlikely. And we may be. But how can we know? All we know is that we are here. Based on the evidence, the chances of us being here are not one over *n*, some number, however larger. They are one out of one. We do not know of any other universes. *Universe* by definition means "all that is." This does not mean that we can't speculate about a vast profusion of unseen worlds. But when we do, it is metaphysics, not physics. The universe is not a repeatable experiment—at least not by us, at least not so far.

A most wonderful example of humanity's befuddlement when it comes to calculating probabilities is the infamous Monty Hall Paradox. Based on the old game show *Let's Make a Deal,* which featured cackling housewives in ridiculous garb engaging in various bargains with the host, Monty Hall, the Paradox, like the show, features three doors. Behind one of the three doors is a shiny new Corvette. Behind the other two are mules. Assuming that you, as a right-thinking technological human being, prefer the automobile to the animal, your task is to choose one of the three doors. There is only one wrinkle: After you choose a door, and for the purposes of this illustration we will assume that you choose Door Number Two, Monty Hall opens one of the other doors, apparently at random, and reveals a mule. Let us say he reveals the mule to be behind Door Number One. Monty Hall then gives you the opportunity to switch doors. So the question becomes, Should you stick with your original choice, in this case Door Number Two, or should you switch to Door Number Three? Or does it even matter?

When I asked my Nobel Prize–winning uncle, Sheldon Glashow, no mathematical slouch, this question, he said, "It doesn't matter."

"That's what most people think," I said.

"Most people don't think," he said.

In fact, as I informed him from my reading, you should always switch: After seeing one door with a mule behind it, the chance that your original choice was wrong has not lessened any, but the evidence of another door being incorrect should induce you to abandon your selection, as the remaining door now is more likely to hide the sports car.

Unless you wish to make a minor contribution to reversing the greenhouse effect by riding a mule, rather than a car, you should switch. Statistically, under these conditions, you should always switch.

Fortunately, the events took place at a seder, and another uncle of mine, Daniel Kleitman, an MIT mathematician, was there to back me up.

Although it seems like it's a one-out-of-two, or fifty–fifty, proposition, the chances of the coveted red Corvette being behind Door Number Three in the Monty Hall Paradox are really two out of three, or 66.66 percent. If you want the car, you should always switch. The reason for the cognitive illusion, the mistaken attribution of probabilities in the Paradox, has to do with the way we frame the question to ourselves once one of the doors is opened. Since only two doors are left, and we see one mule, we naturally assume the probabilities are one out of two. It is as if all the information we need is right before us, in the form of the three doors—now two, because one is opened—on the stage. Although Monty Hall, a thinking agent, is right there, he is the moderator, whom we assume to be neutral, and therefore innocent.

In fact, we should frame not only what we see on stage but also Monty's brain and the information it contains. For Monty Hall *always* chooses a door with a mule behind it, and then asks you if you want to switch. If he didn't, there would be no Paradox: He would show you the gleaming red car, and the experiment would be aborted right there. But this never happens. In other words, Monty only apparently opens a door at random. The chances of your original pick being correct are one out of three, or 33 percent. These chances do not change when a door with a mule is displayed. Statistically, the chance that you chose wrong originally has not changed. Now that you have more information—another wrong choice has conveniently been thrown your way—you should always switch. The 66 (actually 66.6666) percent chance that the car is behind one of the two doors you didn't pick is now, because of Monty Hall's opening of the other door, concentrated on the door you've neither picked nor seen.

The best intuitive way to understand the true probabilities behind this cognitive illusion is to imagine a *Let's Make a Deal* stage with one hundred doors, behind ninety-nine of which are mules. Say you choose Door Number Twenty-three. Monty Hall then opens Doors One through Twenty-two, showing all mules. He leaves Twenty-three closed because it is your choice, then opens Doors Twenty-four through Sixty-eight, leaving Door Sixty-nine closed. Finally, he opens Doors Seventy through One Hundred. Behind all the open doors, ninety-eight of them now, are mules. Your decision is: Should you stick with your original choice, Door Twenty-three, or should you switch now to Sixty-nine?

In this case you would have to be more stubborn to stick with your original, random selection. You have now seen almost all the doors open except one. The chances that your original choice was correct are 1 percent and that you were wrong 99 percent. All those other doors being opened has not changed that. You have been given the opportunity to switch to the door that has a 99 percent chance of covering a gleaming sports car. You should choose Door Number Sixty-nine. That is where the Corvette is, and the moderator has been kind enough to show you every other possible wrong choice except your own. Moreover, if you had gotten lucky and chosen the car right away, there would be no Paradox; the game would be over from the start.

Aside from my uncle, one of the many smart skeptics who were originally fooled by the Monty Hall Paradox was my father. With no mathematician uncle in attendance to back me up, he remained skeptical even after the explanations given above. I can't say I was disappointed or surprised. I once joked to my brother that my father, who was a head taller than his parents, owed his superhero-like status—his physical stature and mental abilities and celebrity—to the effect of chemicals and radiation near his parents' home in Rahway, not far from the noxious New Jersey Turnpike. Most parents seem like superheroes to their young children.

My mathematically minded brother Jeremy, a music software developer, also got the Monty Hall Paradox wrong. Which is not meant to be a critique, only a further example of the overwhelming power of this cognitive illusion to fool. And Jeremy's reaction was illustrative of another human foible. For he said not that it doesn't matter, but that you should always stick with your original choice. Once we make a choice, we feel a sense of identification with, attachment to it. It is *our* choice. We are more reluctant to let it go. The common saying that we use to pat ourselves on the back for such egoism is that we have the courage of our convictions. But as Nietzsche said, he had the courage to go *against* his convictions. Changing one's mind is considered a sign of mental weakness. But changing one's mind in light of additional evidence is a sign of mental flexibility, and science.

Mathematicians now agree that you should always switch. Your original choice had a 33 percent chance of being right. When Monty opens the other wrong door, your chances of having guessed right originally don't magically jump up to 50 percent. They remain the original one out of three. Thus, two out of three times when you switch—abandoning your ego-invested original decision—you will get the shiny red Corvette.

The Paradox is only the tip of the iceberg of academia's growing recognition that humans are prey to cognitive and probabilistic illusions. This is irksome and unwelcome in a way, as we use our instinctive intuition of probability and logic to guide us, and to give us the right answer when we realize we can't necessarily trust our chances. For example, when we see a woman levitate on stage—or on YouTube, for that matter—we figure (rightly in this case) that it is a magic trick. The chances of a woman really levitating are dwarfed by the chances that our senses have been tricked by a professional prestidigitator.

A whole new wing of economics, called behavioral finance, has even arisen that highlights the illogical decisions people make based on their faulty sense of probability. Some of its practitioners have enjoyed a Nobel Prize in Economics. It turns out that our judgment of probabilities is biased by our emotional system, which favors the avoidance of the pain of loss over the pleasure of gain. This has important consequences in personal finances. Avoidance of loss prevents people from making money. Given a choice, for example, the average person would rather take a sure win of $150 than a one-out-of-four chance of winning $700—even though the average win in the second scenario is larger, $175. On the other hand, if it's a question of losing, people would rather risk a larger loss—say, a one-out-of-four chance of losing $700, for an average loss of $175—rather than admit defeat and give up a sure $150. This human prejudice also affects entire cultures. The rise of risk sharing via stock and commodities markets arguably favors societies that have instituted systems to counterbalance the natural human bias against risk taking. Although failure is avoided by not taking risks, so is success. As the late physicist Philip Morrison once told me, an expert is not somebody who is always right. An expert is someone who makes all possible mistakes.

If Nobel prize–winning scientists, gifted mathematicians, and advanced computer programmers can overestimate—thinking they have a 50 percent chance rather than a 33 percent chance—the probability of their being right, what does that say about the rest of us? To me it suggests that our personal and cultural estimations of the probable correctness of our own beliefs may well be wildly overblown. We do not know how much we do not know. Even wildly popular opinions may be completely wrong. This is especially the case when we stodgily dig our trenches and defend our original views in the face of mounting counterevidence.

I find it amusing that creationists, drawing on selected bits of scientific theory, argue that the universe is so statistically improbable that it must necessarily have been made by a willful creator God. Would an all-powerful being that could create anything make, in his own image, beings who push forth their young from between their urethra and anus? Would a benevolent God allow children to be abused, gazelles to be ripped apart by jackals, and corrupt politicians to gain ascendancy?

Most likely not. As Aldous Huxley pointed out in *The Perennial Philosophy,* rather than us being made in his image, the traditional Judeo-Christian-Islamic deity seems to be made in our image—specifically in the image of the sort of petty, patriarchal Middle

Eastern tyrant who ruled society during the time and place when the Bible was written. Such potentates did not appeal to logic and love so much as they ruled by fear and intimidation. An emotionally unstable Ruler of the Universe who instills fear, throws temper tantrums in the form of weather and natural catastrophes, and punishes those who do not worship him seems an unlikely candidate for creator of all that is.

Nonetheless, the culturally familiar metaphysical doctrine that an all-powerful being created the universe has slipped into popular science under the name of the anthropic principle. This principle, which suggests that the universe is bizarrely perfect for the emergence of human beings, points out that slight variations in any number of a host of natural laws and constants would have produced humanless universes. For example if gravity were just slightly different, planets—and the life that exists on them—might not be stable in orbit. If the gravitational coupling constant were ten times larger, there would only be blue giant stars that would burn too quickly for human life as we know it to evolve; if it were ten times smaller, then the universe would be full of red dwarf stars not hot enough to keep water liquid at our orbital distance.

There are endless arguments like this. Protons inside atoms, although more massive than orbiting electrons, have precisely the same electrical charge; if they didn't, atoms would be unstable. If the strong nuclear force coupling constant were slightly weaker, only hydrogen would exist, and other atoms, such as carbon, phosphorus, and oxygen—crucial for life as we know it—would be absent. An infinite number of minute variations would lead to universes without stars, or planets, or the elements needed for life, or time spans necessary for intelligent life as we know it from the fossil record to evolve.

Here, some would say, is evidence for God. Creationists may say it blatantly. But it is also latent in the assumptions and conclusions of certain physicists—sometimes funded by foundations that would like nothing more than to find a scientific proof of God—claiming to be merely calculating the probabilities in an unbiased way. If their calculations came out in a way to suggest that, as Voltaire lampooned Leibniz in *Candide,* this is the best of all possible worlds, then such physicists would not be so likely to get funding from organizations that want not just to believe, but also to scientifically prove, that there is something deeply special about human beings. For example, is it possible for these motivated physicists to use their mathematical prowess to imagine a universe whose constants are so arranged that organisms such as ourselves are not occasionally the victim of severe back and tooth pain? Probably, but that silly metaphysical exercise, suggesting that this is not the best of all possible worlds, and therefore that as an example of divine creation it is glaringly wanting, would not be telling certain people what they devoutly want to hear.

However improbable this universe and our appearance in it, we can specify a still-more-improbable universe that is both more realistic and less a testimony to an omnipotent deity. For example, the nexus of universal constants, initial conditions, and laws of nature that allows the production of our precise universe of hydrogen atoms, stars, planets, and so on, is admittedly extremely unlikely. But still more unlikely—and by the implicit reasoning of the anthropic principle, more divine—would be a slightly different and therefore more improbable universe in which Neil Armstrong not only travels to the moon but sips Cointreau-spiked Tang on the way. Any detail, however bizarre or crude, theoretically makes the world more improbable than it would otherwise be and thus more persuasive evidence of God and his miraculous handiwork. However, many more detailed universes can be imagined, and they are not necessarily more full of grace, beauty, and love than this already imperfect one. The logic of the anthropic principle can be used to suggest that there are an infinite number of universes ranging from slightly to highly superior to this one. That they are not observed to be this one is hardly testimony to the covert presence of a benevolent, omnipotent creator.

Examining magic tricks that give the impression of radical improbability, I argue that the physicists' and creationists' arguments that life is too unlikely to have arisen by chance are just wrong. Number crunching as to the possibility of alternative physical laws and constants is not science but a speculative exercise. That is not to say that it should not be done but rather that it should be recognized for what it is, metaphysics. The seeming bizarre impossibility of our being conscious would not be noticed were we not conscious. This selection bias means that any number of other universes could exist that were not conscious of their prosaic lack of awareness. The universe's startling awareness of itself, through us, may seem—and be—infinitely surprising. But despite the appearances of scientific propriety, the anthropic principle is metaphysics. As statistics, it is fundamentally flawed, since its sample size is one. (A sample size of thirty is often considered to be the minimum for statistical significance.)

Interestingly, a card trick sheds light on this problem. The great Canadian cardsharp Dai Vernon published it under the name "The Trick That Cannot Be Explained." Although not requiring much, if any, sleight of hand, the trick, an excellent one, remarkably simple in method, has fooled both laymen and magicians. (Dai Vernon, known as "The Professor," famously fooled Harry Houdini with a different trick in which a card, placed second from the top, rises to the top.)

The Trick That Cannot Be Explained and our universe are connected because in neither case—the card trick nor our universe—do observers know what to expect. As with the Monty Hall Paradox, the frame, the stage, may have far more than meets the eye, or mind. Anthropic principle speculations upon the alleged improbability of the cosmic conditions leading to man are just that, speculations: We have only one universe and thus, unlike in the case of the Gaia hypothesis, where we actually have other planets with which to compare the improbability of the physical environment of Earth, all alternative cases remain hypothetical. Which is to say that the anthropic principle is metaphysics presented as physics. Now, there is nothing wrong with metaphysics in the sense of the philosophical assumptions governing our physics. But they need to be recognized for what they are, logical inferences and hypotheses that remain beyond the ken of science proper because they cannot in principle be tested. This all said, magic can help us to understand our situation. The strangeness of a universe in which humans exist suffers or benefits from the weirdness of an immensely improbable selection bias: All the other universes, which we cannot see but can metaphysically posit, that do not give rise to or possess observers will never be known, will never be "conscious of" themselves. Despite the mystery of consciousness, the absence of consciousness removes any potential audience to marvel at it. This fact combines with the mathematical regularity of the natural laws of our universe and our ability to imagine infinity to suggest that ours may be one example of an infinite number of unseen cosmos. These facts also must be considered in connection with the notable quirkiness and arbitrariness of our form of life—which, far from suggesting creative perfections, suggest one of an infinite number of unseen versions. The final piece in this metaphysical puzzle is the mathematical fact that the formula to generate an infinite series, $1+n$, is simpler than the formula to specify any finite series. Thus, at least mathematically, an infinite number is simpler and more elegant than a finite version.

The Trick That Cannot Be Explained provides a real-world reality check with which we can compare the metaphysical wonder of this cosmos of which we have only one example. In it, the magician—not God, but, originally, dapper, gray-mustached Dai Vernon—removes (say) a matchbook and, in plain view of the audience, but without letting them see what he is writing, makes his "prediction." He then slides the paper under a glass or somewhere else where there is no chance it can be toyed with or touched by him. A volunteer is then presented with a pack of cards, which she is asked to shuffle and cut several times. This part of the trick (unknown to the audience) may vary. Sometimes after a couple of shuffles, the top card is turned faceup, and it is seen to match the prediction. At other times the deck may be spread across the tablecloth in a ribbon and the volunteer asked to draw out one card—say, the 2♥—which is again seen to match the prediction: Without touching the matchbook, the prestidigitator asks the selector to read what is written there and, sure enough, it is the 2♥.

Impossible, you say? First, realize that this effect is in principle similar to that which is "explained" (really not explained, just described) by the anthropic principle: It is one of extreme unlikelihood. Now, one out of fifty-two, remarkable as it may be, cannot compete with the unlikelihood of our cosmos in some of the metaphysical speculations of motivated physicists calculating the possibilities of our being here. Nonetheless, the effect is of the same basic type—good luck bordering on the miraculous and suggesting supernatural involvement of some sort.

But the secret is devastatingly simple. It hinges upon the fact that the volunteer neither knows what to expect nor sees the alternative realities marked by distinct outcomes. At any point during the course of the shuffling and cutting—which because there is no knowledge that the trick has begun in earnest yet is assumed to be merely preparatory—the magician may notice and present (if it comes up) the 2♥. At which point the trick ends, and the matchbook prediction is read off. If the chosen card does not appear during the course of the volunteer's initial shuffling or after her first cut, she is asked to shuffle and cut again. The more this goes on, the greater the chances of the "random" card popping up. Time is on the magician's side.

After a few shuffles, if necessary, the cards are spread and the volunteer asked to touch one. Of course, even at this late stage (although the volunteer does not know how "late" it is, not having seen alternative versions of the effect) in the trick, if she touches the right card, the trick seems even more miraculous—the previous shuffling and cutting being implicitly dismissed as preparatory. Here is selection bias, our human dismissal of events we do not find meaningful and our attachment to and wonder at events we do find coincidental. Of course the volunteer in the red dress may not choose the right card, but then she hasn't been told to select, only to touch, a card. The wrong card (although again, the spectator doesn't know it's wrong, because she has no idea the trick has even started), turned facedown, will be used as a pointer. "Insert it anywhere," Vernon will say. Again, she may push it right next to the "miraculously found" 2♥—another minor miracle. If not, she will certainly place it somewhere in the deck, say next to a five that is five away from the "chosen" card. One way or the other Vernon is, to paraphrase Debby Harry of the 1980s glam punk band Blondie, gonna get you.

Like the seeming miraculousness of our universe, this trick works because the spectator has never seen it before. There is nothing to compare it with. If she could see the wider context, she would not be so amazed. She would see that the series in which she discovers the chosen card is not the sole one leading to the glittering jewel of extreme coincidence, but one among many. When these many are factored in, the trick becomes inevitable rather than impossible. I submit that, like us and the "trick" of our bizarre but regular cosmos, we are missing a big—huge, in fact infinite—picture. There need be no conscious agent behind it attempting to trick us. But the source of our wonder may be similar: seeing only one outcome and being amazed at its unlikelihood—not because it is in itself unlikely, but because we have not seen all the other outcomes.

Metaphysically considered, the wider context, the other outcomes that we do not see, makes us "misoverestimate" the unlikelihood of this world in which we find ourselves—a world that, marvelously, contains us; but that, not so marvelously, is far too ramshackle and arbitrary to be the work of a competent all-powerful deity heaven-bent on producing the miracle of a single universe. One way or another Dai will persuade our understandably gullible volunteer that she has chosen the 2♥. Because neither she nor the rest of the audience has any experience of this card trick ever being done differently, it will appear to be quite remarkable—one out of fifty-two or, if the joker and extra joker are included in the pack, one out of fifty-four.

We are also gullible, it seems to me, and in a similar position to the volunteer befuddled by seemingly impossible unlikelihood. But with the knowledge of how easily human beings are fooled when it comes to estimating probabilities, we can assume that we are not privy to a vast bevy of alternatives. Some—perhaps an infinite number—of these unseen alternatives, if we could experience them, might seem more miraculous than our own. And many—perhaps an infinite number—would seem less.

If we are an infinitesimal

fraction of an infinite number of unseen universes, our own form, which may seem miraculous but is far from perfect, is more easily explained. The alleged miracle of human existence would be more profound if humans weren't such flawed, imperfect, and confused beings. As

Stephen Hawking writes in *Black Holes and Baby Universes and Other Essays,* "I do not agree with the view that the universe is a mystery. . . . I feel that this view does not do justice to the scientific revolution that was started almost four hundred years ago by Galileo and carried on by Newton. They showed that at least some areas of the universe . . . are governed by precise mathematical laws. Over the years since then, we have extended the work of Galileo and Newton. . . . We now have mathematical laws that govern everything we normally experience."

Why would a cosmos created specifically by a perfect being show such shoddy craftsmanship as bunions, bladder infections, and body hair? The alleged unlikelihood of our existence is betrayed by the fact that we are here. Moreover, whatever the improbability of physical constants and other quantities to produce a universe with stars and life, it is easy to calculate that the chances of a universe arising with stars and life and bipedal humans such as Gomer Pyle and Jack the Ripper, with Hostess Twinkies and an Italian statue of Virgin Mary in disrepair that, as I read, fell down and killed one of the devout, is *even more unlikely.* What are the chances? They are vanishingly small? But they are smaller in this less desirable world with unfortunate women who miscarry their love children, die in childbirth, and give rise to the progeny of rapists than they are in a slightly more divine world where women don't suffer so much, more men are gentlemen, and beautiful white tigers and whales with brains bigger than ours are not in danger of extinction.

Humans die.

Frog species go extinct. Galaxies explode. Our life here is geologically a brief one. Perhaps we can take solace in the relative immortality of our existence as part of something bigger than us. But in the long run our lives here in the Holocene are a flicker, a shadow passing more or less gracefully over the candle of the universe. When we die the elements of our bodies return to the biosphere, entering new organisms, until those organisms die, and so on, until the sun explodes, an event astronomers predict, on the basis of nuclear physics, to occur some five billion years hence.

By which time a lot could have happened.

If it hasn't happened already and we, limited creatures that we are, aren't late to the party. As mentioned earlier, the light from the stars that reaches our eyes often lasts much longer than the stars themselves. The more distant the stars, the greater the discrepancy. Everything in life may be like this, seemingly there but really gone—or from another, still more encompassing point of view, seemingly gone but really here. We do not know but that all of what we see, or seem, is but a reflection of a larger indivisible eternal whole.

Science writer John Horgan captures this vertigo well in a mini essay titled "Fractal Free Fall":

> Mac, my 10-year-old son, comes home from school conveying a classmate's speculations: What if the whole universe is just a germ in the stomach of a giant? And that giant's whole universe is just a germ in an even bigger giant's stomach? And *his* universe? The old conundrum has him thrilled, a bit anxious: How far up and down does everything go? Scientists, the good ones, never cease to be piqued by our betwixt-and-between predicament. Gazing into deep space, astronomers imagine our cosmos embedded in a vastly larger multiverse consisting of countless worlds, each ruled by its own distinct physics. Giants galore! Looking in the other direction, however, some physicists discern a fundamental increment beyond which no smaller thing exists. This is the Planck scale, 10^{-35} meters, where superstrings supposedly shimmy quarks, protozoans, zebras, nuns, quasars—this whole weird show—into being. Just a guess, of course, a wishful one, meant to ward off the vertigo induced by our apprehension of bottomlessness. The Mandelbrot set, mother of all fractals, may be more apt a metaphor for our plight. Magnify the set's crazy-lace border, and you keep spiraling down into new realms, iterations of the primal pattern, forever. No matter how far you burrow into the heart of things, you never arrive. There is no ground of being.

A "fractal" is an irregular-appearing geometric figure that, upon closer examination, exhibits self-similarity: that is, it enfolds its entire pattern into each tiny fragment. The overall fractal repeats itself in itself at various scales of analysis. Although an exploration of the properties and usefulness of fractals has been made possible by the introduction of high-speed computers and graphics programs, the idea of the fractal, of a part that contains the whole, is universal. Is God a fractal, each of whose parts—in a geometry we can glean but not fully comprehend—contains the whole? Buddhism's Net of Indra, an enormous net above the Palace of Indra studded with pearls, each of which not only reflects the whole but also, in a sense, *is* the whole may be an apt metaphor for our vast and connected reality. The universe in David Bohm's quantum theory is holographic, an "implicate order" in which seemingly separate things and particles really have no existence independent of the whole. A striking image from a Hindu text describes the mouth of the baby god Krishna, inside of which can be seen an entire solar system complete with sun, planets, and stars. In Jorge Borges's story "Aleph" a man finds, by a stairwell in a dingy basement in Buenos Aires, a small iridescent sphere that seems to be spinning but really contains the entire world, all space, and everything in it shown infinitely from all angles. Jesus Christ, in the Gnostic Gospel of Thomas (and in Luke), teaches that Heaven is "inside" us. Tendai, a sect of Japanese Buddhism, suggests that the whole and its parts interpenetrate each other—that a dust grain contains Buddhas without number and that the entire cosmos exists within a strand of hair.

To see a world in a grain of sand,
And a heaven in a wild flower,
Hold infinity in the palm of your hand,
And eternity in an hour . . .

The poison of the honey bee
Is the artist's jealousy. . . .

A truth that's told with bad intent
Beats all the lies you can invent. . . .

God appears, and God is light,
To those poor souls who dwell in night;
But does a human form display
To those who dwell in realms of day. . . .
—William Blake

Sure it sounds
like science fiction. But in comparison with James Gardner's idea of intelligent life being the secret fulcrum of the reproducing cosmos, or the utopians on Huxley.net thinking it will be possible in the future to combine psychopharmacology and nanotechnology to find a cure for pain, I believe the idea that we, despite our finitude, are fractally fixed within an infinite universe makes sense. It seems impossible that we could imagine infinity if the universe weren't in some way, in some dimension, infinite. And because we are connected to it we, too, are, at least in that dimension, infinite.

Interviewed by the admiring creator of the first rock-and-roll magazine, *Crawdaddy,* Philip Dick confided in journalist Paul Williams, then writing for *Rolling Stone,* that the most amazing thing in the universe was our imagination:

> After having sent my ship of inquiry out into the universe—you know, to find out the mysteries of the universe—the ship comes back after twenty six years and says the greatest mystery in the universe is . . . [unintelligible]. And then I say, "What is the answer to the mystery?" And it says, "Well, I just said, it's the greatest mystery."
>
> And all I have learned is *where* the greatest mystery in the universe is located. It's between my left ear and my right ear. It's just an incredible thing. It's like, I mean, I know, I've read, this is true, that there are more possible connections between the neural circuitry of the human brain than there are stars in the universe. That should tip you off right there.

Is the structure of the biosphere itself not in some sense "fractal"? Is it a matter of biological identity—of organization as organism—cropping up at progressively more inclusive levels, from cell, to animal, to technological biospheres—themselves global replicas, planetary ecosystems in miniature? Sophisticated fractal graphics—patterns that repeat their overall structure in their minutest parts—became possible only with the aid of the modern high-speed computer, able to perform the same operation many times to generate intriguing, detailed designs. Perhaps the reproduction of the biosphere similarly emerges from a simple operation performed a great many times: the replication of organisms within the crowded environment they are continually altering. And this replication, in turn, although highly faithful and chemically fascinating, is but a variation on a thermodynamic theme. The repetition of living structures, including the technology that has now become part of our lives, ensures the continuing use of energy and production of entropy in accord with the second law.

The machines that have helped us grow so fast are also affecting the biosphere as a whole, in particular Gaia's ability to keep herself cool. Cosmically speaking, this cannot be considered abnormal, but rather reflects the tension between the benefits accruing to fast growers who tap into energy sources and the advantages to more stable, slowly growing forms. The first are like shooting stars that burn out in seconds; the second, slow-burning candles that stay lit for hours.

It is as if life, a kind of slow-burning fire, a chemical reaction, has caught ablaze in a corner of the universe. The question, not so much for us as for our descendants and life as a whole, is whether this fire will catch and spread to other areas of the universe. But this is a temporal question. The fact that we are alive and connected to the universe suggests that the universe itself is already, at least in part, alive. Inchoate, it is perhaps developing to become more conscious, evolving toward a form of absolute understanding, total information awareness that we can scarcely imagine. If human beings testify to something inherent in its essential nature, then our capacity for perception and reflection may be only the slightest foretaste of what is to come. If God did not exist in the past, perhaps his equivalent will evolve in the future.

I've spoken about the future but I've also pointed out that, technically the future never arrives. I believe another possible avenue of the Todorovian is time. Linear time is necessary for causality to occur, and yet linear time may be illusory. Einstein said as much, and the thermodynamicist Ludwig Boltzmann argued that the human sense of time going forward might relate to our position in space. Novelist Paul West speaks of "the immensity of the now." Or as Janis Joplin said, "It's all the same fucking day, man." What this suggests from a narrative perspective is that the future and past are ideas rather than realities and that, in fact, the here and now, the present, contains all that is, was, or will be. We think of time in terms of space: We speak, for instance, of the "distant" past. But to a being not tethered by linear time, narrative itself would be possible, as the whole story would already be there—hatched in its entirety, with no drama or purpose, with nothing to unfold. The story, desire, the movement toward an end, death, that defines us vanish in the absence of the linearity of time. So does causality, and the scientific quest to see how things relate to one another. Hinduism and mystic Christianity, among others, have posited that the Godhead falls into time and flesh from another realm. This realm is not simply the science-fiction future realm of a Heaven or Hell or somewhere else we go to after we die, but a realm that is itself atemporal and eternal. A place outside of time as we normally understand it. A land of already always. We may imagine that ultimate reality—God or the Godhead, call it what you will—does this to entertain itself, since already always existing is not as entertaining as manifesting in animal form and pretending that your time is limited. It seems logical that something doesn't come from nothing. And since we perceive something, something must have always already existed. Being born into the world may be like a magic trick in which a concealing veil is stripped away. We are finite beings but the changing matter that makes our bodies, and perhaps minds, may have always been here. If matter and consciousness are eternal, they don't need to violate the logic of magic technique and appear from nothing. Their covering would just somehow have to be stripped away to reveal their core. In fact, according to philosopher Heidegger, the Greek sense of time was like this: What appears is not new, but preexisting in a frame where the partition covering the future is stripped away, revealing it at present. The Hopis saw the future as behind them, because it was invisible, while we picture the past as behind us, because that is where our footprints are when we are following a path. Both ways of seeing are viable, although diametrically opposed.

Such metaphysical speculations, cultural relativism, and linguistic interpretations, as Philip Dick brilliantly understood, can be made literal in science fiction. In "Prominent Author" he takes us into the home of a businessman who works for a conglomerate corporation in the future. The firm, beta-testing a teleportation device, enlists a low-level worker—one of Dick's typical average-Joe protagonists—for the job. During the testing period the businessman notices a tear in the wall of the tunnel-like contraption that instantaneously transports him to work. One day he crouches and, to his surprise, sees tiny people on a hillside through the tear. They regard him as the giant face of an awesome being. Without his boss's knowledge, he arranges with a friend in another wing of the corporate offices to use, on the sly, automated translation machinery. With this he translates the message from the tiny humanoid aliens. When his boss finds out he has breached security on the beta-test, the worker is fired. During his scolding he is informed of something rather startling: It turns out the language the aliens communicated in was not so alien, after all. It was, rather, ancient Hebrew. Somehow the rift in the teleportation device had put him in contact not only with another part of the cosmos, but with Earth during a former time. And the inquisitive messages this average man sent against the directions of his bosses were more than accepted by the other civilization. They were revered as holy writ. Indeed, the "Prominent Author" of the story has, by a bizarre feedback loop, created the historical miracles of encounters with God reported in the Bible. His giant size relative to the ancient Hebrews is explained by the expansion of the universe over the intervening centuries. Thus our ordinary man is downsized from his corporation even as he is "upsized" through Dick's literal interpretation of the meaning of God in a recursive universe. No longer needing to commute, he hangs around the house admiring his library, which contains—as he boasts to a neighbor (and to his wife's consternation)—a few pretty decent books.

Dick's story perhaps illustrates Spinoza, the meeting of the finite and the infinite. For Spinoza, God is best accessed not with a priest or book, but directly, by looking within ourselves to the mathematical regularities of logic and rationality. God—or what might be reverse-acronymized from Grand Overmind Design—has an infinite number of infinite attributes. Consciousness and space are two that are accessible to us. G.O.D. may not be completely outside of the world, as imagined in Judeo-Christianity, nor inside the world playing a sort of metaphysical joke of hide-and-seek on itself, as imagined in Hinduism. For Spinoza, whose theology inspired Einstein, the real world and ultimate reality overlapped but were not identical. Moreover, because time, as accessed by the mathematical imagination, has no obvious arrow demarcating present from past, the real time of the Godhead is free of partition into past, present, and future. Eternal, like the laws of science, ultimate reality does not choose or actively interfere with "creation" for the simple reason that things like choosing and planning occur within finite and linear human time. For G.O.D. everything already is. As Spinoza pointed out, it is a more noble design that creates everything for once, thereby bypassing the annoying need for divine intervention, personal discussions, and tinkering with perfection. But of course *design* is probably not the right word, since such a universe already always is.

The idea of the eternal present outside of time's would-be flow suggests that the future garden of universal consciousness that the cosmos could become after sufficient evolution may already be here. The future and the past are already here in this line of type, this paper-thin sliver of the Earth spread between your hands. Although numerous philosophers and peoples could now be cited, let us take, more or less at random, the comments of German philosopher and co-inventor (along with Isaac Newton) of the differential calculus, Gottfried Leibniz.

Leibniz saw the universe as an infinitely detailed image of itself. He was encouraged by his mathematical successes, once defining a

Perhaps this is the only possible universe in which everything and its opposite are simultaneously true.

straight line as "a curve, any part of which is similar to the whole, and it alone has this property, not only among curves but among sets." He believed that minute portions of the world are as precisely organized and as complex as large portions, and that the "connection of all created things with every single one of them and their adaptation to every single one, as well as the connection and adaptation of every single thing to all others, has the result that every single substance stands in relations which express all the others. Whence every single substance is a perpetual living mirror of the universe."

A more up-to-date formulation of this microcosmic view of a correspondence between the individual "I" and the universal "all" can be found in many of the writings of physicist David Bohm, who, in trying to find the common ground between relativity theory and quantum mechanics, has proposed that a person is in some sense a microcosm of the universe; therefore what we are is a clue to what the universe is. In Bohm's admittedly speculative view humans are not isolated from but are enfolded with the fabric of the universe. Looking within we may find not only fantasies unrelated to "the real world" but traces of, and corollaries to, all that is. Our internal apprehension of the infinite is not a random fluke. We are fractal reflectors within the glittering fabric of reality. We are fragments of the infinite.

In Bohm's notion of the implicate order—that everything is enfolded within, or implies, everything else—the totality of existence is enfolded within each region of space and time. Although the universe may come unstitched, each part of it is like a seed that can, properly sown, give rise to the whole. "Everything implicates everything in an order of undivided wholeness." One of Bohm's favorite examples of this idea is a transparent container full of very viscous fluid akin to molasses in which ink is dropped and stirred very slowly. If you watch ink stirred this way, it unwinds into a gray thread. But if the stirrer is reversed, the droplet is reconstituted—it becomes "explicate," in Bohm's language. In Bohm's view each part of the universe, each person, is altered, jumbled, enfolded—scrambled like the gray thread. And yet preserved in each one of us are traces of the unity of the whole—past, present, and future.

In 1922 George Perrigo Conger submitted a thin book in partial fulfillment of the requirements of a PhD in philosophy at Columbia University. The monograph was on the history of microcosmic theories in Western philosophy. At that time he summed up the chances for the future of theories comparing the universe outside to the world within. Microcosmic theories, Conger wrote, are

> philosophical perennials [that] decline when the problem of knowledge is made a difficulty, as well as when interest in the supernatural on the one hand or the humanistic on the other upsets the balance between a consideration more evenly divided between man and the universe; and the theories are likely to be suspected or forgotten in a period when the data of the sciences accumulate faster than they can be organized.

At the time, the appearance of an evolutionary microcosmic theory

> [did] not seem very likely, because the prestige of the natural sciences is now so overwhelming; but if the time ever comes it is possible that the successive complicated repetitions of pattern according to some microcosmic theory may, like the chords of a Pythagorean lyre, or the recurrent motif of a Schopenhauerian symphony, appeal to imaginations eager to catch a strain of what the ancients felt to be the cosmic harmonies. This is, as was said, a field practically unexplored; but certainly somewhere in it there is something which has gone out of the modern world.

In the decades since this was written we recognize that the time is now ripe for a microcosmic evolutionary theory of the universe. Individually, each of us has two sides. One faces inward, to our minds and the cells that compose us. The other faces outward, to life and the universe beyond. We are substructures within larger social systems consisting of members of our own species and, still more important, elements within multispecies assemblages running recycling ecosystems. The cells within us, like the ecosystems of which we are a (currently unstable) part, share similar organizational structures. In both the systems that compose us and the global processes in which we are becoming an increasingly important part, matter cycles and energy flows.

The ancients looked up and saw in the night skies signs of a "large man." This was not just a passing metaphor but a whole microcosmic theory of a correspondence between the little world of the human individual and the great one of the universe that encompassed him. Looking forward, it is possible to imagine a scenario in which the cosmos becomes animated in a way our intellectual forerunners and midnight stargazers may never have imagined: If life continues to unfold in the direction set down here—with individuality reestablishing itself at ever-greater levels—the universe may come alive. And yet perhaps it already is. Death becomes—death already always is—the illusion of life's absence.

Our destiny is to traverse the universe. Yet that is exactly what we are doing, right now.

We should have known something like this would happen.

For life is a wave and, the more it changes, the more it stays the same.

All movement, change, and growth is error.
All determinate form is error.
The movement toward sexual union, although as movement in space illusory, is not a form of error.
The desire apparently completed by sexual union is a sign of unconditional form.
Only the unconditional is not illusory.
You are conditional.
The universe is conditional.
You-universe is an aspect of the unconditional.
This note is an aspect of the unconditional.
This has been a note from the Holocene.
Over.

AFTERWORD
TWELVE MYSTERIES

In this book I've addressed some of the deepest questions we can ask—questions sometimes not even asked by religion. But instead of repeating pat answers, I've thought them through from scratch, attempting to give answers by using science, logic, and my experience as a sleight-of-hand magician.

- Why does life exist?
- Why do we drink water?
- Can we save the Earth from global warming?
- Are human beings central and special?
- Is it possible that we've arisen by pure chance?
- Is the Earth an organism?
- Are we part of its exobrain?
- If Earth is alive, can it reproduce?
- Can the universe?
- What does the future hold in store for us?
- Does God exist?
- What is the nature of ultimate reality?

My answers have been tentative but straightforward.

Why does life exist?

I have said that life exists to spread energy in accord with the second law of thermodynamics. The process we call life is unusual but natural, a growing system that cycles matter in a region of energy flow. Global society's focus on energy is a reflection of life's natural energetic purpose.

Why do we drink water?

We drink water because life arose in water some three and a half billion years ago. Atomically water is mostly hydrogen, as is the universe itself. The human body is more than 70 percent water, including the human brain, heart, blood, sweat, and tears. Hydrogen, the lightest element, escapes to space when not kept around by gravity. Composed of water, swimming in it, and raiding it for its hydrogen to combine with carbon and other elements to make its watery cells, tenacious life may be the reason why our planet is anomalously wet. Although Earth's gravity is puny compared with that of Jupiter and Saturn, water, necessary not only for life but for oceans and plate tectonics as well, has been retained here because it is deeply involved in global life's material cycling. The local abundance of hydrogen over the deuterium found in heavy water, and more common in space, suggests that life has kept not just water around on our planet but water of the kind found when it evolved on a hydrogen-rich planet relatively lacking in deuterium—hydrogen's isotope, which is exactly like it except that it has a neutron as well as a proton in its atomic nucleus.

Not only is life composed mostly of water, but early on, more than two billion years ago, judging by radioactive dating of the deposition of oxidized elements in Earth's crust, bacteria mutated to split water molecules into their constituent hydrogen and oxygen atoms. Made mostly of hydrogen, but formerly getting it directly from the atmosphere or from hydrogen sulfide from volcanoes, evolving life now took it from water itself. The result of this photosynthetic trick was the release of oxygen—useful to us but highly reactive and still toxic to many forms of bacteria—into the atmosphere. Today our atmosphere, unlike the almost completely carbon dioxide atmospheres of our planetary neighbors Mars and Venus, is about one-fifth free oxygen gas.

Can we save the Earth from global warming?

Like other thermodynamic systems, global life likes it cool because surface coolness is coincident with the spread of entropy, mostly as heat. Ice cores recording the last four hundred thousand years clearly show increasing levels of carbon dioxide in our atmosphere coincident with a trend of increasing global mean temperature. A vast consensus of peer-reviewed climate science papers in the last thirty years documents global warming and its probable link to fossil fuel emissions.

James Lovelock argues that nuclear energy is the only viable alternative to meet global society's energy needs. Jessica Whiteside argues that millions of tons of carbon dioxide might be removed quickly from the atmosphere by nuking basalts. The result would be an acceleration of the natural process of weathering, producing limestone technologically instead of geologically as carbon, forced by an air pressure gradient down into the basalt, instantaneously reacted with that mineral's exposed calcium and magnesium to remove carbon from the atmosphere, thereby potentially reversing the greenhouse effect.

Are human beings central and special?

No: We are *like* a virgin or we, adopting the immortal words of Tallulah Bankhead spoken in reference to herself, "are as pure as the driven slush." Human beings share traits not only with other living beings but also with inanimate matter-cycling, energy-transforming, entropy-producing systems, one type of which became capable of tapping into energy continuously by reproducing. Like a young woman seduced by the compliments—then disillusioned by the rampant carousing—of her charming unfaithful wooer, we have to realize we are not that special. Realistically, we belong to a class of three-dimensional energy-transforming systems that become more organized locally as they add heat and waste to the environment. The history of life is marked by the battle between accessing energy and materials to keep the energetically favored system going, and the wear and tear and destruction of such systems as they expand, evolve, lay waste to the environment, and devour one another. The history of science does much to deconstruct the wobbly human ego. Copernicus showed that Earth moves around the sun, chemists that life is formed of the most common of elements, and Darwinism that we descend from symbiotic, gene-swapping bacteria, cannibalistic microbes, slimy fish, four-legged animals, and hairy apes.

Is it possible that we've arisen by pure chance?

The universe gives us a sample size of one. This is insufficient as the basis of statistical conclusions. Although there is no direct evidence for other universes, the cosmos in which we find ourselves begins to make sense if we consider it one of an infinite number of others that we do not perceive. Not only is it easier algorithmically to generate an infinite than a finite number, but we must be honest with ourselves and note not only the startling fact of our own awareness but also the many quirks and absurdities of our lives here on Earth. If we were the sole prize of a divine artificer, we would no doubt not be riddled with imperfections such as bad knees, toothaches, and suicidal tendencies. The mathematical ability to imagine infinity suggests that it may not be just a theoretical fancy, but rather a real attribute of a universe that, as science has shown, is largely unperceivable to us.

Is the Earth an organism?

Earth's surface, full of living things, resembles an organism. Like a tornado reducing an air pressure gradient, or an autocatalytic chemical reaction reducing a chemical gradient, Gaia—James Lovelock's name for Earth's surface as a single living being—measurably reduces the solar gradient, meaning the temperature difference between the sun at 5,800 degrees and outer space, with a temperature of 2.7 degrees Kelvin. It has been pointed out that, unlike familiar organisms, Gaia reuses its own materials. Gaia may thus be better regarded as a global ecosystem or, better, a macroorganism. Not only is there strong geological evidence for this macroorganism maintaining its temperature within rather narrow boundaries despite external perturbations (the sun's luminosity has increased), but the concentration of reactive oxygen in the atmosphere has not changed much in hundreds of millions of years. If Earth's surface were a simple physical system, oxygen would react with hydrogen compounds—it would burn up and come to a steady state. That it doesn't suggests that Earth's surface is alive.

Are we part of its exobrain?

In the 1920s the religious Pierre Teilhard de Chardin and the scientific Vladimir I. Vernadsky developed, each in his own way, the idea of a noosphere—a sphere of thought and technology growing from the biosphere. In fact, mind-like processes are all around us. Without any brain or cells, a tornado "figures out" how to choreograph an immense number of air molecules in such a way that they eliminate an air pressure difference. Our cultural focus on tapping energy to continue and expand our existence may not be so different. We—with our Internet, global telecommunications satellites, televisions, radios, videos, and ever-growing forms of transmitting signals—are not alone but part of a thermodynamic system in which increasingly sedentary humans appear increasingly similar to signal-happy planetary neurons.

If Earth is alive, can it reproduce?

The formation of enclosed bioshelters, experimental recycling systems, glass-covered microbial ecosystems, and human efforts at housing an ecologically complementary suite of life-forms in closed containers—necessary for long space voyages—suggests that the biosphere has already begun to make copies of itself. Although human beings presently seem necessary for this production of new, miniaturized ecosystems, that does not invalidate the claim that the biosphere itself is on the verge of reproduction. In order to be pollinated, some orchids require myopic wasps to mistake their fragrant parts for willing mates. Although they do not make themselves, farm implements from hoes to high-tech tractors are involved in cyclical systems of agricultural production that feed people, make farmers money, and lead to the production of more farm implements. As Samuel Butler pointed out in the nineteenth century, "What is a reproductive system, if it be not a system for reproduction?" The same may be said for the biosphere. The export of recycling stations to satellites or the colonization of Mars would be de facto reproduction of Gaia. Technology may ultimately be seen not as a human phenomenon but as part of the development of a viable means of biospheric reproduction.

Can the universe reproduce?

Lee Smolin posits that new universes with varying laws and constants pop out on the other side of black holes. Other things being equal, universes giving rise to more black holes will be more successful, suggesting that our universe is the partial result of a mother universe that produced a lot of black holes. Noting that unstoppable technology is moving to the point at which black holes will someday be produced, James Gardner suggests that intelligent life may itself be part of a cosmic creation process. In this view we are part of the progeny of a universe making more of itself via the evolution of life, intelligence, and black-hole-producing technology. I find this view intriguing because it posits a deep natural role for intelligent life, usually portrayed as anomalous or divine. But I also find it is suspicious because it brings back the scientifically debunked position of humanity as central, this time not geographically central but functionally so. This seems to me possible but unlikely. The function of life to cycle matter, use energy, and produce more entropy than would occur without it, by contrast, is a demonstrable fact. The difference is that intelligent life in the first instance is considered to have a unique function while, in the second its function is shared with all known materially cycling systems.

What does the future hold in store for us?

Based on the evolution of other species, we can expect to become either extinct or pseudo-extinct. Species said to become pseudo-extinct disappear from the fossil record not because they become extinct but because they branch off into offspring species. We may hope that human beings, perhaps through our technology and intelligence, will avoid the typical fate of a backboned species, to become extinct after a few million years—which we've already had! A few forms in life's history have hit upon solutions that have ensured their massive representation in future life-forms. Perhaps the most striking example are the ancestors to mitochondria. These oxygen-using bacteria, by teaming up with larger cells, have helped propel the evolution of surface life-forms such as plants, fungi, and animals. Human beings, too, perhaps by applying our technology to serve as protective bubbles for extraterrestrially recycling life, may avoid extinction by teaming up with other organisms in new collectives.

Does God exist?

Depending on how we define the word, he may. A personal God that made humans in his likeness to testify to his glory seems unlikely, as humans are so imperfect. A God that, as my father says, has a gray beard, counts the hair on everybody's head, and tallies the fall of every sparrow seems pettily human. Humorist Jack Handey has remarked that God is said to live inside everybody; I hope he likes enchiladas. The orthodox Christian God, compared with the glorious regularities and universal laws uncovered by science, is simply embarrassing. For Spinoza, God, outside of time, is the universe working according to its eternal laws and much more besides. He is infinite in extent, and we can apprehend another aspect of his infinitude in our consciousness. This sort of God, coincident with the timeless laws of nature, is what Einstein believed in. Such a God does not act capriciously. He does not stoop to divine intervention, which would speak more to the ceaseless tampering of an unsuccessful artist than the meta-perfect genius of an eternal being. Hopefully, when we look within, we discover more than taco sauce.

What is the nature of ultimate reality?

Ultimate reality may be refractory to our attempts to describe it in words. However, according to writer, philosopher, and science-fiction author Aldous Huxley (who took LSD as he was dying), the great religions and religious figures were all talking about the same thing. This was a timeless zone that was not personal. In both *advaita* in Hinduism and *annata* in Buddhism the individual and the universe are separate only through a sort of illusion, sometimes called *maya.* Interestingly, computer metaphors from contemporary neuroscience tend to support this view: Ordinary consciousness is not all there is but only a tiny subdivision, a screen that filters an otherwise-too-confusing reality and allows us to go on in our cloistered way. Just as your computer has a user-friendly operating system, we are normally graced with a user-friendly consciousness. Drugs or epilepsy may disrupt this screen and give us a deeper—if more frightening and confusing—view. This view, accessed by mystics and described in sacred texts, tends to poetically describe the dissolution of the ego and to reference a timeless realm, a realm where time is seen "end-on." Logically, since nothing cannot give rise to something, the facts that matter and consciousness exist suggest that they always existed.

If matter and consciousness are eternal, then our perception of them as temporary may be mistaken. It is interesting to contemplate that only a universe separated from itself can perceive itself as an observer perceives what is observed. But if observer and observed are ultimately one, then the separateness is illusory.

You and the universe are the same thing. You are the universe looking at itself and you and the universe mutually implicate each other. Jesus in the Gospel of Luke also teaches that Heaven is inside us. And a perhaps better translation of Jesus' words in Matthew 5:48 is "Be inclusive, as your father in Heaven is inclusive" (rather than "Be ye therefore perfect, as your father in Heaven is perfect"). As the mystics say, "All is one," or "Thou art that." Philip K. Dick identifies the essential quality of humanity as our ability to empathize—but this is ultimately a nonquality, as it connects us to everything we are not. The atoms of our bodies return to the biosphere, itself in space. In the fullness of time we do not exist as humans but as part of a perhaps endless universe that will produce new beings who will also be able to say "I" and dwell temporarily in the illusion of separateness. In short, we must deceive ourselves to perceive at all. Free of the illusion of separateness, ultimate reality (putting it in the negative terms favored by mystics) may thus be neither disturbed nor enchanted by knowledge of its own existence. It is impersonal, undivided, and eternal. Nonetheless I am assuming it is partly accessible to us since we are part of it. Ultimate reality may also have a fractal structure in which the whole is contained (if "through a glass darkly") in each part.

Mystery Bonus Question: Where do socks go when they die?

Sometimes they cling to sheets, sometimes they're still in the dryer, and sometimes there's no accounting for their disappearance. Okay, I know I said *twelve mysteries* but this last one, a minor, less metaphysical bonus, seems worth mentioning. A recent personal discovery persuaded me I may have found at least a partial answer to this enduring conundrum—the place you would least expect to look: the dryer itself. I was cursing my luck that I had yet again lost a sock, rendering its would-be partner an ostracized mismatch at best, when lo and behold I found the sneaky garment upon the inner spoke-like fin of the collaborating clothes dryer. There certainly must be many other repositories containing missing members of abandoned sock pairs, but for some reason this humble discovery—of the missing sock in the dryer all along—seemed significant. It suggested that what we think is lost is really, whether we are clever enough to tease out its existence or not, right there in our midst, if not in plain view or on our very person. As comedian Stephen Wright says in his trademark depressed monotone, "You can't have everything. Where would you put it?"

SOURCES

Earth

Blanchot epigraph: Blanchot, Maurice. "Literature and the Right to Death." In P. Adams Sitney, editor, *The Gaze of Orpheus and Other Literary Essays,* translated by Lydia Davis (Barrytown, NY: Station Hill Press, 1981), 46.

Baidu questions: "What the Chinese Want to Know." *Foreign Policy,* http://blog.foreignpolicy.com/node/3076 (a blog by the editors), January 12, 2007.

The very word biosphere . . .: Corning, Peter A. *The Synergism Hypothesis: A Theory of Progressive Evolution* (New York: McGraw-Hill, 1983).

Derrida, Jacques, and Maurizio Ferraris. *A Taste for the Secret* (Cambridge, England: Polity Press, 2001).

Sagan, Carl. *The Varieties of Scientific Experience: A Personal View of the Search for God* (New York: Penguin Press, 2006).

The most influential environmental photograph ever taken: Galen Rowell, cited in "100 Photographs That Changed the World," *The Digital Journalist* and *Life* magazine, http://digitaljournalist.org/issue0309/lm11.html.

Daisy World: Lovelock, James. "Geophysiology: A New Look at Earth Science." *Bulletin of the American Meterological Society* 67, no. 4 (April 1986): 392–397. For a less technical treatment, see Lovelock, James. "Gaia: The World as a Living Organism." *New Scientist* (December 1986): 25–28. A simple flash animation presentation can be found at: http://library.thinkquest.org/C003763/flash/gaia1.htm

Relationship between plankton and global cooling: See Monastersky, Richard. "The Plankton–Climate Connection." *Science News* 132, no. 23 (December 5, 1987): 362–365.

Nature *article:* Schwartz, Stephen E. "Are Global Cloud Albedo and Climate Controlled by Marine Phytoplankton?" *Nature* 336, no. 1 (December 1988): 441–445.

Current status of Gaia: Schneider, Stephen H., James R. Miller, Eileen Crist, and Penelope J. Boston, editors. *Scientists Debate Gaia: The Next Century* (Cambridge, MA: MIT Press, 2004).

Westbroek, Peter. *Life as a Geological Force: Dynamics of the Earth* (New York: Norton, 1992).

Cooling effect of aerosols from industry masking global warming: Lovelock, James. *The Revenge of Gaia: Earth's Climate Crisis and the Fate of Humanity* (New York: Basic Books, 2006).

Zoologist Richard Dawkins disputes evolution at any level beyond that of the individual (but what is an individual if not an evolved group of cells?): in Dawkins, Richard. *The Extended Phenotype*. (San Francisco: W. H. Freeman, 1982).

Thomas, Lewis. "Beyond the Moon's Horizon—Our Home." *New York Times* (July 15, 1989): 25.

Don Anderson quote: Lovelock, James. *Ages of Gaia: A Biography of Our Living Earth* (New York: Norton, 1988), 105.

Borges, Jorge Luis. *The Book of Imaginary Beings* (Harmondsworth, Middlesex, England: Penguin Books, 1980), 21–22.

Hutton lecture: Delivered by James Hutton in 1785 before the Royal Society of Edinburgh. Cited in Lovelock, James. "Gaia: An Example of Large-Scale Biological Design." Presented at the Conference on Biology as a Basis of Design, Perugia, Italy, 1988.

[I]f you have life on a planet . . .: From the television program "Goddess of the Earth," *Nova,* transcript no. 1302, originally broadcast on PBS, January 28, 1986, WGBH Educational Foundation.

Lovelock, James. *Gaia: A New Look at Life on Earth* (New York: Oxford University Press, 1979).

Lem, Stanislaw. *Solaris,* translated by Joanna Kilmartin and Steve Cox (Fort Washington, PA: Harvest Books, 2002). And check out the very amusing official Web site: www.lem.pl.

Eddington, A. S. *The Nature of the Physical World* (the Gifford Lectures, 1927) (Cambridge, UK: Cambridge University Press, 1933), 74–75.

Klauder, John R., editor. *Magic Without Magic: John Archibald Wheeler: A Collection of Essays in Honor of His Sixtieth Birthday* (San Francisco: W. H. Freeman, 1972).

The important Haldanian precept that the universe is stranger than we can imagine: Haldane, J. B. S. (John Burdon Sanderson), *Possible Worlds and Other Essays*, (London: Chatto and Windus, 1927), 286.

Jeans, Sir James. *Physics and Philosophy* (Cambridge, UK: Cambridge University Press, 1948).

Cordyceps on the Web: http://www.mushroomexpert.com/cordyceps_militaris.html

Terry Carr on marketing the Bible as science fiction: Carrère, Emmanuel. *I Am Alive and You Are Dead: A Journey into the Mind of Philip K. Dick,* translated by Timothy Bent (New York: Metropolitan Books, Henry Holt, 2004), 16.

Water

Rumi epigraph: Parabola, the Magazine of Myth and Tradition (spring 1988): 136.

Broecker, Wallace S. *How to Build a Habitable Planet* (New York: Columbia University Trustees, 1998).

Relatively enriched levels of deuterium: Press Release 00-19, *American Geophysical Union News,* www.agu.org/sci_soc/prrl/prrl0019.html, June 27, 2000.

"We Can Remember It for You Wholesale": In Dick, Philip K. *Selected Stories of Philip K. Dick* (New York: Pantheon, 2002).

Cyanobacteria evolution releasing oxygen from water: Margulis, Lynn, and Dorion Sagan. *Microcosmos: Four Billion Years of Microbial Evolution,* foreword by Lewis Thomas and updated authors' preface (Berkley: University of California Press, 1997).

For more on writing as a pharmakon, see "Plato's Pharmacy" in Derrida, Jacques, *Dissemination*, 1972, Barbara Johnson, translator, (Chicago: University of Chicago Press, 1981), 63-171.

An animation of the formation of the moon from debris after Theia crashed into the Earth thirty-four million years after its origin: can be found at http://en.wikipedia.org/wiki/Giant_impact_hypothesis.

Human-caused extinctions in the Holocene: See, for example, Leakey, Richard E., and Roger Lewin. *The Sixth Extinction: Patterns of Life and the Future of Humankind* (New York: Anchor Books, 1996).

I am grateful to Nathan Currier for permission to quote from his personal letter dated July 9, 2006, to Earth system scientist Tyler Volk.

Hayes, Brian. "Goodness, Gracious, Great Balls of Gaia!" (a review of *The Revenge of Gaia*), *American Scientist* (online edition), www.americanscientist.org/template/BookReviewTypeDetail/assetid/53119;jsessionid=aaaff1vtzpuTAZ, 2006.

Tarnas, Richard. *Cosmos and Psyche: Intimations of a New World View* (New York: Viking, 2006), 19, 21, 24–25, 491–492.

Astronaut Eugene Cernan: Cited in Frank White, *The Overview Effect: Space Exploration and Human Evolution* (Boston: Houghton Mifflin, 1987), 206.

Look again at that dot . . .: Sagan, Carl. Lecture at Cornell University, October 13, 1994, www.bigskyastroclub.org/pale_blue_dot.html.

It is important for the human race . . .: Stephen Hawking quote from Sylvia Hui, "Hawking: Humans Must Colonize Space," Associated Press, www.space.com/news/060613_ap_hawking_space.html, June 13, 2006.

almost as if you have come back from the future . . .: Eugene Cernan (the last man to walk on the moon) cited in White, *The Overview Effect.*

Living matter is a specific kind of rock . . .: Lapo, Andrei Vitalyevich. *Traces of Bygone Biospheres,* revised from the 1979 Russian edition (Moscow: Mir Publishers, 1982), 58.

Turnover times of atoms, cells, and tissues in the human body: Spalding, K. L., R. D. Bhardwaj, B. A. Buchholz, H. Druid, and J. Frisen. "Retrospective Birth Dating of Cells in Humans." *Cell* 122, no. 1 (2005): 133–143.

Water, you have neither taste . . .: Antoine de Saint-Exupéry, cited in Lapo, *Traces of Bygone Biospheres,* 34.

Art does not deliberate . . .: McKeon, Richard, editor. *The Basic Works of Aristotle* II.8 (Princeton, NJ: Princeton University Press, 2001), 251.

Don't you see . . .: Giordano Bruno, cited in Krumbein, Wolfgang E., and Betsey Dexter Dyer. "This Planet Is Alive: Weathering and Biology, a Multi-Faceted Problem." In J. I. Drever, editor, *The Chemistry of Weathering* (Boston: D. Reidel Publishing, 1985), 145.

Air

Brodsky epigraph: Brodsky, Joseph. *So Forth: Poems* (New York: Farrar Straus Giroux, 1998).

Merriam-Webster definition of spirit: www.m-w.com/dictionary/spirit.

In the particle-picture . . .: Jeans, Sir James. *Physics and Philosophy* (Cambridge, UK: Cambridge University Press, 1948), 204.

For more on consciousness as an interface: See Nørretranders, Tor. *The User Illusion: Cutting Consciousness Down to Size* (New York: Penguin, 1999).

Abram, David. "Merleau-Ponty and the Voice of the Earth." Lecture at the annual gathering of the Merleau-Ponty Circle (New York: New School for Social Research, 1983).

It sometimes seems that I must work not only for myself . . . and *The right of freedom . . .:* Cited in R. K. Balandin, *Vladimir Vernadsky,* translated by Alexander Repyev (Moscow: Mir Publishers, 1982), 25, 47–48.

One thing seems to be foreign . . .: Cited in Lapo, *Traces of Bygone Biospheres,* 18.

. . . biosphere . . . common parlance: Vernadsky, cited in M. M. Kamshilov, *Evolution of the Biosphere* (Moscow: Mir Publishers, 1976), 78.

Grinevald, Jacques. "Sketch for a History of the Idea of the Biosphere." Presented at the International Symposium on Gaia, Theory, Practice, and Implications, organized by the Wadebridge Ecological Centre and ECORPA at the Worthyvale Manor Conference Center, Camelford, Cornwall, England, October 21–27, 1987.

The biosphere is a as much . . .: Vernadsky, cited in Balandin, *Vladimir Vernadsky,* 47–48.

. . . green burning: Ibid., 103. Vernadsky may have borrowed this phrase from English clergyman and chemist Joseph Priestley (1733–1804) or from French chemist Antoine Laurent Lavoisier (1743–1794). Vernadsky's two biospheric principles can be found in Lapo, *Traces of Bygone Biospheres,* as well as in Vernadsky's *The Biosphere,* an abridged version based on the French edition (London: Synergetic Press, 1986), 79–80.

Carruthers's calculation: Carruthers, G. T. "Locusts in the Red Sea." *Nature* 41 (1890): 153.

. . . anecdotal . . .: Lovelock, James *The Revenge of Gaia: Why the Earth Is Fighting Back—and How We Can Still Save Humanity* (London: Allen Lane/Penguin, 2006), 160.

vague, imprecise word . . .: Ibid., 161.

Ulam, Stanislaw. "Tribute to John von Neumann." *Bulletin of the American Mathematical Society* 64, no. 3 (May 1958): 1–49.

Folsome, Clair. "Microbes." In T. P. Snyder, editor, *The Biosphere Catalogue* (Fort Worth: Synergetic Press, 1985), 51–56.

McHarg, Ian. *Design with Nature* (New York: Doubleday, 1967), 43–54.

The Bios project: Ivanov, B., and O. Zubareva. "To Mars and Back Again on Board." *Soviet Life* (April 1985): 22–25.

Kamshilov experiment: Kamshilov, *Evolution of the Biosphere,* 91–93. Original experiments described in M. M. Kamshilov, "The Buffer Action of a Biological System," *Zhurnal Obshchei Biologii* 34, no. 2 (1973) (in Russian).

Sagan, Dorion. *Biospheres: Reproducing Planet Earth* (New York: Bantam Books, 1990).

Rusty Schweickart cited in Sagan, Dorion. Biosphere II: Meeting ground for ecology and technology. *The Environmentalist* 7(No. 4): 271-281.1987,

Allen, John P. "Historical Overview of the Biosphere 2 Project," www.biospheres.com/pubhistoricaloverview.html, 1990.

Hydrick interview: "Confessions of a Leading Psychic," www.unexplainable.net/artman/publish/article_2685.shtml, November 24, 2003.

James Hydrick YouTube video demonstration of psychokinesis (telekinesis) on That's Incredible: www.youtube.com/watch?v=niKlbt-L8HU.

Jung, Carl Gustav, cited in Tarnas, *Cosmos and Psyche,* 51–52.

Jung, Carl Gustav. *Synchronicity: An Acausal Connecting Principle*. In *Collected Works of Carl Gustav Jung,* translated by R. F. C. Hull, edited by H. Read, M. Fordham, G. Adler, and W. McGuire. Bollingen Series XX (Princeton, NJ: Princeton University Press, 1953–1979, volume 8, 1952).

Russell, Peter. *The Global Brain* (Los Angeles: J. P. Tardier, 1983), 82.

... stumbled, if not exactly by mistake ...: Jeans, Sir James. *The Mysterious Universe* (New York: Macmillan, 1930), 4.

An omnipotent creator ...: Ibid., 11–12.

Fire

Pi Yen Lu epigraph: Rothenberg, David, translator. *Blue Cliff Record: Zen Echoes* (New Paltz, NY: Codhill Press, 2001), 46.

George Carlin quote on life going backwards: http://thinkexist.com/quotation/the_most_unfair_thing_about_life_is_the_way_it/345791.html.

Birch, Charles. "Why Aren't We Zombies?" In John Cobb, editor, *Neo-Darwinism and Process Thought* (Grand Rapids, MI: Eerdmans, 2007 [in press]).

"Fear": You can flip through a facsimile of an old book containing this short story by Guy de Maupaussant at www.openlibrary.org/details/secondoddnumbert00mauprich.

Sullivan, Kevin. "$25 Million Offered in Climate Challenge: Tycoon Hopes to Spur Milestone Research," *Washington Post* Foreign Service, A13, www.washingtonpost.com/wp-dyn/content/article/2007/02/09/AR2007020900693.html?nav=emailpage, February 10, 2007.

France, Vermont, Connecticut nuclear reactors; National Cancer Institute study; and Nuclear Energy Institute X-ray comparison: Sudhakar, Nina. "Eye on Nukes." *New England Watershed* (spring 2007), 66–68.

Butler, Samuel. *The Works of Samuel Butler,* volume 20: *Note-Books* (New York: AMS Press, 1968).

Gardner, James. *Biocosm: The New Scientific Theory of Evolution: Intelligent Life Is the Architect of the Universe* (Makawao, HI: Inner Ocean Publishing Company, 2003).

I do not agree that the universe is mysterious ...: Stephen Hawking, cited in Clifford Pickover, *Laws of Nature* (Oxford: Oxford University Press, forthcoming).

Hawking, Stephen, *Black Holes and Baby Universes and Other Essays* (New York: Bantam, 1994).

Mac, my 10-year-old son . . ..: Horgan, John. In Bill Atkinson et al., *Within the Stone: Nature's Abstract Art* (San Francisco: BrownTrout Publishers, 2004), 30.

After having sent my ship of inquiry. . . ..: Philip Dick quote transcribed from *The Gospel According to Dick* CD.

. . . a curve, any part of which is . . ..: Gottfried Leibniz, cited in Benoit Mandelbrot, *The Fractal Geometry of Nature* (New York: W. H. Freeman, 1977), 419.

Leibniz, Gottfried. *Monadology and Other Philosophical Essays,* translated by Paul Schrecker and Anne Martin Schrecker (Indianapolis: Bobbs-Merrill, 1965), 156–157.

Everything implicates . . ..: Bohm, David. *Wholeness and the Implicate Order* (New York: Routledge and Kegan Paul, 1980).

Conger, George Perrigo. *Theories of Macrocosms and Microcosms in the History of Philosophy* (New York: Columbia University Press, 1922), 136.

INDEX